AF835448

WIND–DIESEL AND WIND AUTONOMOUS ENERGY SYSTEMS

Proceedings of a contractors' meeting on wind demonstration projects, organised by the Commission of the European Communities, Directorate-General for Energy, held in Mykonos, Greece, 25–26 April 1988

WIND–DIESEL AND WIND AUTONOMOUS ENERGY SYSTEMS

Edited by

H. NACFAIRE

Commission of the European Communities, Brussels, Belgium

Taylor & Francis
Taylor & Francis Group
LONDON AND NEW YORK

Published by Taylor & Francis
2 Park Square, Milton Park, Abingdon, Oxon, OX14 4RN
711 Third Avenue, New York, NY 10017, USA

Routledge is an imprint of the Taylor & Francis Group, an informa business

First issued in paperback 2011

British Library Cataloguing in Publication Data

Wind–diesel and wind autonomous energy systems
1. Electricity supply. Large wind turbine generators
I. Nacfaire, H.
621.31'2136

ISBN 978-1-85166-338-5 (hbk)
ISBN 978-0-415-51586-3 (pbk)

Library of Congress CIP data applied for

Publication arrangements by Commission of the European Communities, Directorate-General Telecommunications, Information Industries and Innovation, Scientific and Technical Communications Service, Luxembourg

EUR 11931

PREFACE

The present publication gives an overview of Community Demonstration projects on "Autonomous and wind-diesel systems" presented during a meeting organised on Mykonos Island, Greece, on 25 and 26 April 1988 for the contractors involved in demonstration projects supported by the Commission of the European Communities ("CEC").

The meeting was held with the collaboration of the Public Power Corporation ("PPC"). It was attended by 55 participants from 9 member states, including representatives of most public utilities of the European Community.

The objectives of the meeting were to:

- assess the state of advancement of the projects, aiming at wind-diesel systems (including projects where wind turbines are connected to a diesel-based grid) and autonomous systems;
- share the experience gained by contractors;
- favorise a possible cooperation between manufacturers;
- inform the utilities of the state of the art;
- stimulate the replication of successful projects;
- visit the wind turbine in Mykonos.

The meeting was opened by Mr A Kravaritis (Deputy Director General of PPC) who referred to the existing plans to install 18 MW of various size WTs within the European Community programmes. Dr M Davis, Director in the Directorate-General for Energy, CEC, welcomed the participants on behalf of the Commission.

Presentations were made on 17 projects. Besides the papers presented on the projects, interesting information was presented on the establishment of wind/diesel plants other than those of CEC projects. There were also presentations by three invited speakers.

The first speaker, Professor N H Lipman, presented a paper on "Overview of wind-diesel activities" The second speaker, Dr G Cramer, presented a paper on "Control and load management systems on wind power plants connected to diesel based grids" and the third speaker, Mr S.E. Andreasen, described the experience gained from the realisation of a wind diesel project in China (a project financed by the Directorate-General for Development of the CEC.

Throughout the meeting, particular attention was given to successful cooperations between utilities, manufacturers and users. The audience's interest was proven by lively discussions with useful exchanges of ideas.

Many valuable and useful conclusions have been drawn, most of which are mentioned at the end of this publication.

I take this opportunity to thank all participants and the Greek organisations for their support and their contributions to the success of this meeting.

H. NACFAIRE
Coordinator for Wind Energy
Demonstration Programme

CONTENTS

OVERVIEW OF WIND/DIESEL SYSTEMS

N.H. Lipman
Head of the Energy Research Unit
Rutherford Appleton Laboratory.

SUMMARY

This overview is made up of three main elements:

1) A look at the main design questions for wind/diesel systems. The criteria affecting design choices for the diesel sets, wind turbines and energy stores are looked at briefly.

2) Examples of current Research and Development and Demonstration projects are given, ranging from small supplies of a few kW to large systems up to 1MW.

3) There is a separate discussion of the largest systems (multi-megawatt) which include a number of diesel sets on an isolated grid, and which permit "multiple-diesel strategies" to be operated.

1. INTRODUCTION

I wish to thank Mr Michael Davis and members of the DGXVII directorate for inviting me to speak at the seminar on "Autonomous Wind/Diesel Systems". My instructions have been very clear and rather daunting: "cover a wide range of systems and scenarios and try to highlight the advantages and disadvantages of each one". Well I'm not sure that I can achieve this ambitious requirement, but I will attempt to cover a fair part of the field and will try to bring out some of the more important arguments.

The first thing that became apparent as I started to address this task is that there is a very extensive world-wide activity in this area. At the EWEA's 1986 Conference "EWEC 86 "(part sponsored by the CEC) there were some 20 papers on wind/diesel R & D. Two recent workshops on the subject took place in May and June 1987, at Dartmouth College near Boston, USA and at Rutherford Appleton Laboratory (RAL) near Oxford, UK, respectively. (The RAL workshop report which contains 15 papers is now available from BWEA). I have drawn on the papers from these 3 events and on the experience of my own research unit in preparing this report.

Let me start by putting the many types of W/D projects into a number of main categories. Even this is a difficult exercise as some projects may combine several different control and operation principles. For example there are projects containing both load control and flywheel storage. Others include "multiple diesel" strategies plus battery storage. Nonetheless, I will define a number of strategic principles. The fact that a "strategy" is required arises from the highly fluctuating nature of the wind; also from rapid changes in power requirements within small electricity networks. Thus a wind power station may be fully supplying an "autonomous" load at one moment and may be in considerable power deficit only seconds later. Hence the need for some magic formula such as wind and diesel. We now know that a simple diesel back-up may not suffice!

Some of the strategies that are being tried out in pilot studies in many parts of the world are listed below in table 1.

Table 1 Wind/diesel strategies

Main Strategy	Subset	Comments	Example
1) Load control	a) At domestic level	a) Frequency Switches	Lundy Isle, UK
	b) Major loads switched	b) Water pumps	Cape Verde(CWD)
	c) Desalination	c) Part day requirement	Fura Ventura (new proposal)
2) Long term store	a) battery	a) allow diesel start/stop	Cape Clear, Eire
	b) pumped hydro	b) allow diesel start/stop	Foula, Scotland
	c) battery/cylic diesel	c) Diesel cycles at max load	Canada
3) Short term store	a) Hydraulic/ pneumatic	a) 5 mins for turbulence	Reading Univ, UK
	b) Flywheel	b) Few mins for turbulence	Imp College/RAL, UK
4) Hybrid systems	a) Load control/ flywheel	a) reduces cycling of load control	Fair Isle, Scotland
	b) Battery store/ flywheel store	b) flywheel for rapid diesel starts	Cape Verde (RISO)
5) Multiple diesel	a) Large Island grids	a) Can optimise loading of diesels	Shetland, Scotland
	b) with storage	b) Improve loading optimisation	Kythnos, Greece

In section 3 I will look at examples of several of these strategies in a little more detail. It should be noted that no two projects are exactly the same. Hence there are very many different strategies being tried out, world-wide, if we choose to look in a little more detail.

2. GENERAL CONSIDERATIONS

Before looking at the specifics of any one strategy I would like to ask some more general questions about what one is trying to achieve in setting up demonstration autonomous wind systems. What are the underlying principles in the design of such systems, and where may we hope to improve on this first cycle of studies and projects.

2.1 Diesel Power Systems, Single or Multiple Diesel

When we talk about "autonomous wind/diesel systems" we may be talking about anything from a 10kW wind turbine operated in conjunction with a 5kW diesel electric set, up to several 1MW single wind turbines being introduced into a 30MW island grid. The problems at these two extremes are so very very different that it is difficult to cover both in one short (30 minutes) paper. The larger systems have several advantages as far as ease of operating strategy is concerned, with respect to very small wind/diesel systems:

a) A large system will have a multiplicity of diesel generating sets, and hence running strategies can be adjusted (stopping or starting diesels) to minimise fuel consumption and to maximise the benefits from the wind power. (however there are limits to this flexibility as large supercharged diesels cannot be started and stopped too frequently or too rapidly.

b) Large loads representing 100's or 1000's of households vary in a much smoother way than very small loads.

c) In the case of multi-megawatt grids we are likely to be introducing a number of wind turbines. If these are geographically separated then they will provide a useful smoothing of wind power fluctuations when compared to a single wind turbine. In short time periods of seconds up to minutes (depending in the degree of separation) the reduction in power fluctuations will go as $\sqrt{N}$, where N is the number of wind turbines (W.T.'s)

For the reasons stated above we need to treat large autonomous diesel grids in a very different manner from the small systems, on which many of us are working.

I define a small system as lying in the size range from 10kW to about 200kW of diesel power. Such a system will probably have only 1 or 2 diesel electric sets and we will be introducing only 1 or 2 W.T.'s into the system. For most of the paper I will be dealing with these smaller systems which have a small number of generating components. The interesting and exciting challenge which we must meet with regard to such systems is how can we devise schemes that can provide very substantial diesel fuel saving (e.g. 30% to 60%), and yet must at the same time be both technically simple and reliable and economically viable. It is interesting at this moment, to stand back a little to examine the merits and demerits of the many schemes that, are currently being tried out in a wide range of imaginative projects.

2.2 The sizing of Diesel Generators

Frequently when we talk about autonomous wind power systems we are dealing with the introduction of wind power to an existing diesel grid. Our first calculations will show how much diesel fuel can be saved when compared to the fuel being burnt in the existing system.

I believe that WE MUST BE VERY CAREFUL here. The original diesel generator may be very much oversized, as is frequently the case. Small diesel systems (e.g. 10kW to 200 kW) are likely to be faced with highly variable and spiky loads. For example the peak load may be as high as 5 times the average load (see Bass and Twidell, 1986). It is not uncommon for the designers of the systems to have chosen a diesel with a capacity of 2 times the peak load (see Dure, 1985), to provide a safely margin and room for future expansion of the load. Such a diesel operation on its own (without wind power) will produce extremely costly electricity. This is apparent if we look at a typical small diesel efficiency curve, as in figure 1. Note that at zero load the diesel still burns about 1/3 of maximum fuel. Furthermore it is not recommended to run diesels for any length of time at small loads and normally a minimum loading of 30% is suggested to avoid bore glazing, oiling up of the silencer and other problems. If this advice is taken seriously and we operated a remote diesel power station at a strict minimum load of 30%, then it would be necessary to incorporate a controllable dump load in the system. Studies (Harrap, 1987) examining the economics of small diesel power supplies suggest electricity costs of the order of 90c/kWh for the situation described above.

Evidently, we would wish to avoid such an extreme situation. Yet in the real world it is not uncommon to find single grossly oversized diesels as in the case above. The practical reality is that there would be no computer controlled dump load, but the operators would probably run the diesel for less than the whole 24 hours, and would endeavour to switch on sufficient loads, whether actually required or not, to give the diesel a reasonable loading.

It is very difficult to quantify the true benefits of a new wind/diesel system which is added (e.g. by way of a "Demonstration Programme") onto a very unsatisfactory diesel network as described above. Yet such unsatisfactory diesel operation is very common in isolated applications around the world.

Now along comes the wind energy engineer with a wind turbine, a storage medium and a micro-processor controller. Recognising that the original diesel generator was grossly oversized he now introduces a smaller diesel which will work in tandem with the wind turbine and the store, to meet the load. Such a situation with and without wind energy has been examined by Harrap (1). He considers a wind/diesel system with battery storage. Whereas, the original diesel is sized at 10x average load, he now replaces this with a diesel sized at 2x average load. One option that he examines is to ignore the wind power and to use the battery/inverter system to improve diesel efficiency. In this way the load demand pattern assumed in the study is met without ever running the original (oversized) diesel generator. The results of this study suggest the cost of electricity produced by the diesel-battery system is likely to be 40% lower than that produced by the diesel-only system. In fact such systems, working along similar lines, are sold commercially in Canada, Australia and elsewhere. They may be referred to as "cyclic charging systems". Usually the diesel is run at full output to recharge the batteries and is then turned off. An alternative mode in which the batteries would be used for peak lopping might also be of interest.

The point that I wish to make here is that the wind/diesel engineer is making a useful contribution to the "isolated community" in devising improved strategies for the operation of the local diesel network. I suggest that he should examine carefully the relative benefits that are achieved by intelligent control strategy of the diesel network and those that are provided by the inclusion of the wind power.

I am not suggesting that the windpower cannot provide considerable economic benefit in its own right. I am sure that it can. But in fairness to the customer we should examine the two stage process:

a) The benefits provided by a newly optimised diesel network with intelligent control.

b) Additional benefits provided by wind power.

2.3 Wind turbine choices and sizing

There are a number of tricky decisions to be made with regard to the design choices for the wind turbine. These include some of the following:

a) Size relative to the diesel set(s) and relative to the load. For high penetration of wind power (e.g. 50%) the wind turbine rating may typically be twice the diesel rating, given that a typical W.T. load factor would be about 30%.

b) Some workers argue in favour of several smaller wind turbines (e.g. 30kW) rather than one larger wind turbine (e.g. 150kW) in a single installation, to make use of the short timescale smoothing effect (see Cramer, 1987).

c) There are arguments both ways in favour of a synchronous generator or induction generator on the wind turbine. The induction generator has the difficulty that it requires an external source of reactive power. On the other hand synchronisation to the diesel grid is much simpler, and dynamic interactions with the grid are less of a problem. The synchronous generator has the advantage of being self-energising, on the other hand it represents a "stiffer" source of AC power and synchronisation and stability problems can be encountered. A majority of projects use induction generators in conjunction with a continually spinning synchronous generator on the A.C. line to provide reactive power. This is usually the generator on one of the diesel sets which disconnects (via a clutch link) from the diesel, when this is stopped.

I would point out that the price that we must pay for adopting such a scheme will be the continuous spinning losses of the synchronous generator, which are quite large, e.g. 700 watts for a 7kW generator (Bleijs, private communication).

d) Finally there is the question of power shedding. If there is excess wind power and the storage is full then wind power must be shed in some way. There are several approaches to this problem.

(i) Some groups make use of a dump load; e.g. the Imperial College/RAL team (see Coonick et al, 1987).

(ii) Another approach is to permit the W.T. to overspeed thus activating a passive pitch control mechanism (see de Bonte and Costa, 1987)

(iii) A third approach is to have a rapid action pitch control on the W.T. This is the method used by Cramer et al. (see Cramer, 1987).

It is not possible to say which of these 3 methods will be most cost effective, without careful examination of the details of each project. Active pitch control tends to be expensive and is not usually favoured for small machines for this reason. Passive pitch control is quite common and well accepted. The power limiting approach of de Bonte and Costa requires additional equipment, namely an AC-DC-AC link, which is also expensive. Yet this scheme does include an additional benefit that some of the excess wind power is converted into additional kinetic energy in the W.T. rotor (which overspeeds). This represents a few seconds of stored energy which is helpful in dealing with short term downward fluctuations of wind power. The dump load approach of Imperial College/RAL also has its advantages and disadvantages. A dump load can be costly, although the design adopted by Imperial/RAL has cut this cost quite a lot. An advantage is that power dumping leads to quite a simple control strategy, and it can also be used to provide a minimum loading of the diesel. This "dumping control philosophy" also permits some of the excess wind power to be used, in practice, in auxiliary loads such as water heaters, etc, as local conditions permit.

2.4 Choice of energy store

There are very many complex issues relating to the several choices of energy storage listed in table 1. I will not attempt to discuss these in any detail in this section, but some points will come up as I discuss individual projects in the final section. Very briefly some of the questions that we must ask are as follows:-

1) Is the energy store providing only a strategic benefit or can it also save on fuel usage in its own right.

2) What is the efficiency of the store (in/out losses), and are the losses dependant on the rate of power flow.

3) Does the store have a continuous standing loss (e.g. continuously spinning machinery).

4) Does the store have a finite lifetime (e.g. batteries will take a certain number of charge/discharge cycles).

5) Does the store have a limit on the charge/discharge rates (e.g. batteries).

6) What are the likely maintenance requirements and costs.

7) Is the system of a complexity that can be handled in very remote areas.

8) Finally and most important, what is the cost of the store including all associated equipment.

Let me make one general remark on storage, as it relates to wind energy, before I go on to the third section of this paper. It is important that we consider the time structure of wind turbine output. If we look at a spectral analysis of wind speed, as presented by Van der Hoven in 1957 (see fig. 2) we see that wind turbulence comes in two main frequency domains. There is a high frequency turbulence (10 to 1000 cycles/hour) which causes us great difficulty in maintaining a short term steadiness of electricity supply. Secondly there is a turbulence in times from about 5 hours to 200 hours, corresponding to movements of weather fronts. This latter is less trouble to us, as it requires only relatively slow control decisions with regard to the operation of the various components of the autonomous electricity system. Those projects that opt for a short term store (e.g. flywheel, hydraulics, etc) are tackling the problem of the high frequency turbulence and are accepting the principle that there will need to be a fair amount of supply switching (wind or diesel, etc) in times of hours. The longer term stores may carry additional benefits of fuel saving, if they can bridge between periods of high and low wind. However, to achieve these benefits the costs must be acceptable and in/out efficiencies sufficiently high. I would be interested to see an analysis showing whether it is cost effective to go to these much longer storage times.

In fairness, I must also challenge the short term storage strategies. These provide the benefit of cutting down start/stop cycling of the diesel to less than once per hour. They also provide bridging power whilst the diesel is being restarted (see Coonick et al, 1987, Slack and Musgrove, 1987 and Bullock and Musgrove, 1987). What remains to be proven is that such systems will be reliable and can be built at a cost which does not spoil the economics of wind/diesel. I believe that the answer to these questions is in the positive, but it is still too early to say, as all of these short term storage projects are still at an early development and demonstration stage.

3. EXAMPLES OF SPECIFIC PROJECTS

In the time available it will not be possible to discuss in detail all of the different types of strategy listed in Table 1. Nor will I be able to do justice to the many excellent projects currently underway in each general category. Nonetheless, I will try to give an impression, if fairly brief, of some of the major initiatives in the wind/diesel area.

3.1.1 Load Control Schemes - Domestic Level

An early approach to autonomous wind diesel strategy was made by Mr Murray Sommerville (see Sommerville and Stevenson, 1984 and 1986) who worked for International Research and Development, and is now a Director of Wind Harvester. He has set up 3 systems, on off-shore islands of U.K and Ireland, in the years 1980 to 1982. These systems on Fair Isle, Lundy and Inis Oirr, differ in some details but are all designed to certain underlying principles.

A diagram of the Fair Isle system is shown in figure 3. A 55kW windmatic design wind turbine is combined with diesels of 50kW and 20kW into a small island electricity network. The basic control strategy is to go for load switching. Mr Sommerville's scheme has 3 priorities of load.

i) The top priority load which is made up of "high quality" domestic loads including lighting and electrical equipment (T.V., radios, hi-fi, etc).

ii) Load group 2 which would be a lower quality uses of electricity, including water heating and storage heaters.

iii) Load group 3 which is essentially a dump load, but can include low priority requirements such as swimming pool heating, etc.

The loads 2 and 3 can be brought-in in a graduated manner, this being accomplished by frequency sensing switches. These switches are installed into every house and are set at a range of different thresholds. The sizing of the wind turbine was such that for much of the time its output was well in excess of the priority load (load 1). Load 2 and possibly load 3 will then take up the excess wind power; the diesel generators being stopped for much of the time. The wind turbine speeds up or slows down slightly as input wind power increases/decreases, the frequency switches bringing loads in or out very rapidly. If the wind power drops too low to meet the priority load (load 1) then one of the diesels is brought on and takes over this load. The supply of the priority load was only guaranteed for 2 periods each day (e.g. 7am to 9am and 4pm to 11pm) although the scheduling may have changed by now. A tariff structure was adopted to encourage the use of wind electricity (when available). As reported in 1983 (see Infield and Puddy, 1983) the tariff structure for Lundy was 7p, 3p and 2p for diesel, wind priority 1, and wind priority 2, respectively.

Progress reports from these projects indicate that they have run very successfully and that wind power has supplied more than 75% of the loads. Let me give a view on the advantages and disadvantages of such an approach, as I have been asked to do.

Advantages	Possible disadvantages
-A simple and robust scheme	-Does frequent switching damage appliances
-No storage medium required	-Tailoring to each specific island may prove too costly in engineers time
-Most of W.T. output utilised	-Not all islands may have sufficient number of low priority loads
-Diesel running greatly reduced	-The low priority loads use electricity to a much lower worth
-Users get a better service than ever before	

3.1.2 Load Control Schemes - Switching of Major Loads

I will look very briefly at the other approach to load control, namely that of using several large lumped loads. Such a scheme has been built by the Dutch CWD group (Consulting Services Wind Energy Developing Countries) at Terefal on Cape Verde and was reported on at the RAL wind/diesel workshop (see de Bonte and Costa, 1987)

A schematic diagram of the system is shown in figure 4 a windturbine has been added to a grid that was originally supplied by a 175kW diesel electric set. This was grossly oversized, as the maximum load was some 45 kW. Electricity supply was provided for two short periods each day as is shown in the load diagram of figure 4b. The original diesel would have been running very inefficiently.

In the CWD scheme a new 70kW diesel generator set was brought in to replace the original oversized set. The wind turbine in the new system is described as "conservatively sized" and is rated at 30kW. Studies at CWD had shown that there was little benefit to having a larger wind turbine, as a diesel start/stop strategy was not to be adopted in this initial pilot project (see fig 4b).

In addition to the original island load, there were 3 additional loads of 11kW each coming from 3 large water pumping stations, each of which had to be run for about 8 hours per day. A much improved load profile is achieved in the new scheme by incorporating the 3 pumps into the island load, as in fig 4d, rather than having them running, rather inefficiently, each on its own diesel (as had been the case previously).

The authors estimated that there would be a saving of just under 1/3 in fuel (i.e. 30,000 litres per year), ignoring load strategies with the water pumps, but this saving becomes some 70,000 litres per year including the water pumps in the operating strategy.

My own conclusion would be that we see again the benefits of load management and a sensible re-optimisation of a remote diesel grid. It would seem to me that the wind energy engineers are having their main impact in achieving this. The benefits from the wind energy, itself. are probably quite small and may not even be economic.

Advantages	Disadvantages/Uncertainties
-A simple approach with real benefits -Control strategy simple to execute -Lumped secondary loads used to good effect	-Continuous running of diesel implies minimum impact of wind power -Wind energy contribution probably about 10% -Would it be simpler to ignore the wind power altogether and concentrate on load management?

3.2 Long Term Storage

Here I will take the case of battery storage in small to medium sized systems. There are many such projects in Europe and around the world. Groups active in the field include Linders et al (1987) at Chalmers University, Sweden, and Lundsager et al (1987) at the Riso test station. Gunther Cramer, who collaborates with Thomas Schott at the DEVLR Test Station in Germany, gave an excellent review of this type of system at the 1987 RAL workshop (see Cramer, 1987 and also Schott et al, 1987). A schematic diagram of the systems that the latter collaboration built and installed on the Irish Island of Cape Clear is shown in figure 5.

A completely new system has been installed including a 72kW diesel electric set, two 30kW wind turbines and a battery store of 100kWh, plus associated two way inverter (120kW). Starting from scratch, and ignoring the old generating system, it has been possible to design a well optimised system. The battery has two purposes:

a) To back up the wind turbines when the diesel is stopped.

b) For peak lopping when the diesel is running on its own and a load power spike in the load exceeds the diesel rated output.

Cramer in his 1987 paper does not indicate what level of fuel savings are to be expected with this newly installed system. However, I would comment that in my judgement the component sizes are well matched. I would expect a possible fuel saving of about 30%. Let me make a few technical comments.

a) This collaboration generally favours two smaller W.T's rather than one larger machine, because of the smoothing effect that can be achieved.

b) They favour active pitch control on the W.T's as a way of controlling output power; hence reducing the need for rapid action dump loads.

c) The diesel set and its synchronous generator can be decoupled by way of a clutch. The generator is kept spinning when the diesel is stopped in order to provide reactive power to the rest of the system. However, this does imply fairly large standing losses. (Most wind/diesel & battery schemes incorporate a similar rotating condenser arrangement.)

d) I note that the battery storage is unusually small (compared to most similar projects) with only 30 minutes of storage at maximum invertor rating. This surprises me as most lead-acid batteries have their lifetime shortened and their storage capacity greatly reduced when charge/discharge times are much shorter than 10 hours (Lucas batteries, private communication, 1988). Perhaps the average power flow in the system is much lower at about 30 kW, in which case the battery storage would correspond to 2 hours (which is still short for lead acid batteries).

Finally I note that elsewhere the authors talk about "rugged and low cost invertors" and of "special batteries". Both of these factors could be very significant, as otherwise the capital cost of a battery invertor system would seem to me to be potentially very high, and might make such systems too expensive for wind/diesel projects, other than those funded by demonstration programmes!

Several years ago in a joint paper with Reading University, we estimated an in/out storage cost for batteries (because of their finite lifetime) of 5p/kWh stored! More recently Michael Harrap has provided me with a figure from his work of 8p/kWh stored. These figures do not include the costs of buying or operating the invertors. Hence I deduce that if a substantial part of the wind power had to be processed in the battery then the cost of "wind electricity" is likely to exceed 10p/kWh.

I have indicated some possible uncertainties, mainly economic, with regard to the battery-inverter approach to wind/diesel systems. I will finish by giving a list of advantages and disadvantages.

Advantages

- Batteries & invertors are a well proven technology
- Additional benefit of peak lopping for diesel operation
- Larger battery systems can bridge between windy and less windy periods

Disadvantages/Uncertainties

- battery & invertor costs may be very high
- batteries have finite lifetime & need some care and maintenance
- Invertors can give harmonic distortion to electricity supply
- some types of invertor can fail in catastrophic manner
- standing losses from "rotating condenser"

3.3 Short term stores

In this area I know of two main activities. The first is work on a hydraulic- pneumatic store which is being carried out by Dr Peter Musgrove team at Reading University. The second is work on flywheel stores, where several teams are active, but where probably the most extensive such programme involves my own team (RAL) working in conjunction with Imperial College and two UK companies (Laing ETE and Hawker Siddeley Power Plant).

3.3.1 The hydraulic-pneumatic store

Two papers at the RAL workshop described this work (Bullock and Musgrove 1987, Slack and Musgrove, 1987). A schematic of the system is shown in figure 6a. This diagram does not show a back-up diesel generator, but there would normally be one in the system. A hydraulic pump/motor is coupled to the electrical power line by way of its own synchronous generator. This is spinning all of the time and may be decoupled from the hydraulic pump/motor at times when this is not in use. The synchronous generator does represent a standing loss, but it can be used as a rotating condenser if an induction generator type W.T. is in the system. Energy storage is in one or several hydraulic-pneumatic accumulators. These devices contain compressed nitrogen gas which acts up on the hydraulic fluid in a separation bladder. Several minutes of storage can be achieved in such systems at an acceptable cost. The storage units available commercially are limited in size (approximately 10-15KW/min) including the back-up storage bottle so that it will often be necessary to stack several in parallel for larger wind/diesel systems. Similarly the pneumatic pump/motors are only available up to a size of 100kW, so again it may be necessary to stack several units in parallel. Hence it would seem to me most likely that this approach will find its best application for the smaller sizes of wind diesel systems, perhaps up to 100kW diesel generator size.

Musgrove and his team have shown that there is considerable benefit to be had from the application of such a short term energy store (see figs 6b and 6c). We see in fig 6b that such a store will bring diesel start/stop cycles down to acceptable levels of about 10 per day. Fuel saving is considerably improved compared to a system with no storage, as we see in figure 6c, graph (a). In fact fuel savings of 50% or 60% are not out of the question for good windy sites. Let me now list the pros and cons as I see them:

Advantages

- hydraulics is a well established technology
- short term store can produce desired benefits
- costs are likely to be acceptable
- application looks good for small systems
- possible use for peak lopping when only diesel is running!

Disadvantages

- hydraulics can be troublesome
- question of maintenance on remote sites
- standing losses from synchronous generator and also from hydraulic motor when running
- in/out losses fairly high e.g. 35% to 60%

Let me finish my remarking that in-out losses are probably not very important for a short term storage strategy as not much of the wind energy is cycled through the energy store.

3.3.2 Flywheel energy store

The flywheel is much older than the steam engine and is used in many applications including the motor car. Two UK companies, Laing ETE, and British Petroleum have been developing higher technology high energy density flywheel stores for special applications. My own experience is with the Laing ETE flywheel which I will discuss in a little more detail. The application to wind/diesel systems is an ongoing programme involving the Energy Research Unit at RAL, the Power Engineering Group under Dr Leon Freris, at Imperial College, and two UK companies, Laing ETE and Hawker Siddeley Power Plant. Apart from a considerable amount of theoretical work, on both system stability and optimisation of logistic control, we also have an experimental programme on a fully operating wind/diesel/flywheel rig. The system is shown schematically in figure 7a and 7b and a photograph of the rig is shown in figure 8.

This system has been operated successfully in a fully autonomous mode since October 1987. A considerable amount of experimental work has been carried out to investigate possible modes of instability and to monitor fuel saving and start/stop cycling. The companies involved are now working on the design of a commercial version of the diesel/flywheel rig which it is hoped can be coupled electrically as an add-on package, to a wide variety of commercial wind turbines.

Our theoretical work leads us to very similar conclusions to those of Musgrove et al. For example our estimates of annual fuel saving are shown in figure 9. We find, as Musgrove does, that fuel savings of 50% to 60% per annum can be achieved, for energy storage of just a few minutes. Our theoretical predictions on start/stop rates are very similar to Musgrove's. (for more details see Coonick et al, 1987 and Lipman et al, 1986).

An important consideration here, as in all cases, must be the standing losses, as the flywheel plus generator are rotating all of the time! In fact the diesel and the flywheel share the one synchronous generator, which minimises losses from this component. (i.e. the generator would be turning anyway when the diesel is running). When the diesel is stopped, the synchronous generator, whilst spinning with the flywheel, will serve a second purpose of acting as a rotating condenser for those cases where the wind turbine has an induction generator. In this respect we will be in a similar situation as regards losses to practically all of the projects discussed in this review paper.

It follows that the only additional standing loss singular to this storage approach will be the losses of the flywheel itself from windage, bearings and gearbox. Windage is very small as the flywheel operates in an evacuated chamber. Bearing and gearbox losses are expected to be small compared to the losses in the synchronous generator.

Finally, it would be fair to comment that although the approach looks very promising, and is likely to be cost effective, flywheels with many kilowatt-hours of storage are not in common use and will need great care in design and manufacture if they are to be reliable over many years of operation. Let me summarise the pros and cons.

Advantages

- -A simple and robust scheme
- -Flywheels are a well known and well tried technology
- -Short term store can produce desired benefits
- -Costs look very promising
- -Applications over a wide range of sizes (kW to Mw)
- -Possible use for peak lopping with respect to diesel
- -Standing losses of flywheel not high
- -in/out efficiency high

Disadvantages/Queries

- -killowatt hours of storage not yet well proven
- -question of maintenance in very remote areas?
- -standing losses from continuously spinning synch. generator (but as in most other projects)
- -simple scheme of fig 7 requires network frequency excursions of ±5Hz

3.4 Hybrid Systems

In this section I will be very brief, simply highlighting the fact that combinations of several of the principles discussed above are now being tried.

3.4.1 Battery storage and flywheel

Per Lundsager has described an advanced wind diesel system which a Danish consortium including Bonus and Riso Laboratory are currently constructing at Santa Caterina on the Cape Verde Islands (see Lundsager, 1987). The outline is quite similar to the German/Irish project on Cape Clear. The battery storage is 66kWh as against the diesel size of 40kW and wind turbine size of 55kW. This means that the batteries provide about some 2 hours storage at typical loads. (still quite a short storage time as applied to lead-acid batteries, it seems to me). As in most other schemes, a continuously spinning synchronous generator is needed a) as spinning condenser then the diesel is not running, b) as the diesel's generator set when the diesel is running. However, as there is a small flywheel attached to the synchronous generator, fast diesel starts can be achieved via an electro-magnetic clutch. Lundsager claims that the flywheel is also helpful in maintaining frequency stability. I will not list pros and cons for this project as the list would be very similar to that for the Cape Clear project.

3.4.2 Load control and flywheel storage

Experiments have been carried out with a British Petroleum flywheel, developed by Dr Rayner Mayer, operated in conjunction with the Fair Isle wind/diesel system. Those involved in the project included Dr C Jefferson of Bristol Polytechnic, Rayner Mayer of BP, and Murray Sommerville of Wind Harvester. The work showed that frequency excursions can be greatly decreased when the flywheel is connected to the network. This in turn implies that the frequency switching of the load will occur less rapidly and starts of the diesel are required less often. In all other respects the advantages and disadvantages listed in 3.1 still apply.

4. MULTIPLE DIESEL GRIDS

Here I am considering a quite different situation. The grids in question are likely to have a maximum rating of between say 5MW and 50 MW. Examples include Scilly Isles and Shetland off the coast of the U.K. and a multitude of Greek and Spanish Islands including Kythnos, Mikonos and certain of the Canary Islands. For islands that have expensive diesel electricity, and at the same time good wind speeds, wind power does look very promising. Pilot projects are currently underway, or at a planning stage for all the islands named above and for many others around the world.

A majority of these early projects are looking at the impact of single, mainly large, wind turbines, addressing a wide range of questions both environmental and technical. But they are not yet putting the most fundamental question to the test, namely, what is the ultimate penetration of wind energy into major island grids that can sensibly be achieved.

Large island grids will normally be supplied by a fairly large number of diesel electric sets, which means that there is usually a fair degree of flexibility of response of this plant to a highly variable wind power input. If a number of wind turbines is involved and these are fairly widely spaced from each other, then windpower fluctuations will be reduced in times of seconds or even of a few minutes, depending on the spacing.

However, the two favourable factors noted above do not mean that major island windpower is without any difficulties. Large diesels require a much longer run-up time to full power (and full efficiency) than the small diesel-electric sets that I have been discussing in the previous sections. For example, the 6MW units on Shetland take some 8 minutes to run up, during which time they changeover from a light fuel oil to a heavier, but low cost fuel oil. Evidently, a very rapid cycling, on and off, of such large diesels would not be efficient or sensible. Hence there are questions of operating strategy to be considered and researched before we can achieve very high wind energy penetration for MW island grids.

I consider most of the projects currently underway as more of TOKEN VALUE than of real significance, with regard to the ultimate technical test of HIGH WIND POWER PENETRATION. However their political significance, in terms of creating a public awareness and hopefully an acceptance of the technology, and their usefulness in terms of understanding local operating problems, are not in doubt.

As regards the very highest levels of wind penetration into large multi-megawatt island grids we must turn for the time being to the theoretical studies. I will look briefly at two such studies:

4.1 Shetland - Multiple Diesel (No storage)

This project is a collaboration between the North of Scotland Hydro-Electric Board, Rutherford Appleton Laboratory and the University of Strathclyde (see Halliday and Gardner, 1987 and Twidell et al, 1987). There are three components to the study:

a) A detailed model of the grid, with windpower added up to very high wind penetrations, to look at operating logistics and potential final saving.

b) A dynamic model to look at questions of transient response and system stability with wind power.

c) Monitoring of two likely wind sites, in advance of the decision to build a CEC funded demonstration 750kW, W.T. on Susetter Hill.

I will say a few words here about (a), the logistic model. A very detailed model has been written by Jim Halliday and Paul Gardner from RAL and the University of Strathclyde, respectively. The model has been written in a very flexible manner so that it can be applied in the future to a wide range of island grid situations. It is of the time-step type and time steps can be chosen from 1 minute upwards. The operation of existing grid plant can be taken into account in a very detailed manner in terms of operating requirements, part load efficiency and minimum and maximum loading of individual diesel sets. The time steps at which advanced decisions will be taken on consigning of plant can also be adjusted to requirements.

Halliday and Gardner have pointed out that such models are only of value if they have been carefully TESTED AND VALIDATED against known operating conditions of real plant. To this end they have made careful comparison with actual Shetland grid operations (with no wind input). This was a very careful and time consuming exercise, to which I was privy, which involved looking carefully at the way the operators run the power station in practice and comparing this with the decision making written into the model. This exercise led to improvement of the model to a point where one could examine the daily operating scheduling of each size of diesel electric set on the system and to compare favourably theoretical prediction with practical reality. Detailed results of this validation exercise are given by Halliday and Gardner (1987). I reproduce here in figure 10 a graph showing monthly operating figures for the groups of diesel generation set sizes, comparison being made between model prediction (M) and real results (R). We see that with the exception of February 1985 (where new plant was first being brought in) there is very good agreement between prediction and real results, giving us some confidence that the model can now be applied to wind plus diesel situations.

Some early results looking at wind on the Shetland grid, up to very high attempted penetrations are shown in figure 11. As usual we see a law of diminishing returns. The 3 curves differs only in the matters of the operating strategy and the extremes of diesel operation that we are prepared to permit (in terms of underload or overload). What becomes clear is that if we must run the large diesel sets conservatively (curve a) then sensible levels of wind-power penetration would be limited to about 15%. Curves (b) and (c) show higher possible levels of wind penetration if we are prepared to run the diesel sets rather more harshly! Further studies are clearly required in which we look rather carefully at what type of treatment large MW diesels can sensibly accept, within a wind/diesel system. It is also apparent that the addition of energy storage and/or load control might be very helpful in aiding high wind penetration.

We will be addressing these questions in the next phase of the project which is due to start in October 1988.

3.5.2 Multiple Diesel and Battery Storage (Kythnos)

Contaxis, Kabouris and Chadjivassiliadis have developed a model which includes battery storage (see Contaxis, 1986). This has been applied to the wind/diesel and battery storage system of Kythnos. This is a much smaller grid than Shetland with a total installed diesel-electric capacity of 1.7 MW. The diesel engines range in size from 100kW to 550kW. Hence operating requirements of the diesels will be quite a lot different from those on Shetland, in terms of run-up times, and permitted operating limits. Nonetheless this is an interesting case of "multiple diesel strategy" at an intermediate size range between the small (100kW) and the very large (tens of MW). The system modelled by this group has 100kW of wind turbines made up of 5x 20 killowatt machines. Battery storage levels from 600 to 2400kWh hours were studied in the model. Results from the model on fuel saving as a function of battery capacity are shown in figure 12. We see that there is practically a linear improvement of fuel saving with battery size up to a capacity of some 1200kWh. There is a further worthwhile improvement up to 1800kWh. The impact of battery storage is also shown in table 2, which I reproduce from their paper.

Table 2 - Summary of results for fuel consumption

	Season 1			Season 2		
	Fuel Savings	Fuel Consumption		Fuel Savings	Fuel Consumption	
Case	Kg	%	Kg	Kg	%	Kg
1	0	0	5085.81	0	0	3291.32
2	1486.09	29.32	3599.71	1561.15	47.43	1730.16
3	2147.95	42.23	2937.83	1703.86	51.76	1587.45

Cases 1, 2 and 3 correspond to diesel only, diesel and wind, diesel and wind and battery storage, respectively Two different seasons are covered in the table. The results for the two seasons are very different, which rather surprises me.

	Season 1	Season 2
Fuel saving (wind)	29.32%	47.43%
Fuel saving (wind + battery)	42.23%	51.76%
Benefit from battery	+12.91%	+4.33%

It appears that the addition of battery storage has a very large impact in season 1 and yet a relatively small impact on season 2, when the initial wind penetrating (no storage) was already very high at 47.43%. I do find these results very surprising, as I would have suspected that batteries would be most helpful in getting from high penetration of wind (like 50%) to even higher penetrations by bridging believers windy and still periods.

One lesson that I glean from these results is that when one has a model working, it is most important to run it on as many cases as possible, in order to gain confidence in the model, and some breadth of insight into the situation that it is describing.

From the results discussed above I would conclude that multiple diesel situations can benefit quite markedly from the addition of a storage medium. This is in contradiction to a commonly held view that says: "once you are dealing with a large multiple diesel grid you have a flexibility of operation that implies that large wind power inputs can be accepted with no need for storage". However, we still need to know whether such a battery storage systems, as considered by Centaxis et al, will be cost effective.

The Imperial College/RAL/Laing collaboration will investigate the application of large flywheels to wind-diesel situations, in a new project which will be starting during 1988.

5. CONCLUSIONS

I have looked at a wide range of projects, many showing a high degree of intelligence and ingenuity, which are currently operating in many parts of the EEC and in the world outside. I repeat that I see all of these projects to be very much of a pioneering nature. It is very important that such "pilot projects" receive support from "Demonstration Funding" or "Aid funding" at this early stage in the development of autonomous wind systems. I believe that a number of the schemes discussed here will prove to be technically viable and reasonably reliable. Different schemes will suit different situations. For example, load control as a solution will suit situations which have a large number of low priority loads, but will not be so effective in other situations with few such loads. The hydraulic storage scheme would seem to apply to fairly small systems (10kW to 100kW) but may not scale higher unless larger storage units and hydraulic motor can be found. Batteries may have a problem with high initial costs and high operating costs, but in areas where electricity costs are high (eg greater than 20p/kWh) batteries could prove most effective as they can make possible very high wind energy penetrations (like 80% or 90%). They can be used to bridge between less windy and more windy periods. The various flywheel schemes look very promising. Very short storage time (a few minutes) can help the strategy of small to medium size wind/diesel systems very greatly. The early estimate of costs for such systems look promising and I believe that they will be able to compete in situations where the "diesel only" electricity cost is as low as 10p/kWh. Larger flywheels with 10 minutes, or longer, of storage are likely to be very helpful when applied to multi-megawatt multiple-diesel grids.

6. ACKNOWLEDGEMENTS

I wish to thank Mr Davis, Mr Nacfaire and Mr Diamantaros of DGXVII for their invitation to present this paper at the Mykonos Workshop. My thanks go also to members of the Rutherford Appleton Laboratory/Imperial College wind/diesel project team for helpful discussions whilst preparing this paper. Finally, special thanks to Mr Michael Harrap, a sabbatical visitor to the RAL Energy Research Unit for many helpful discussions on the subject.

7. REFERENCES

Harrap MJ, "Some aspects of the design of Hybrid Wind Diesel System", Proceedings 'Solar '87", ANZSES, Canberra, Australia. November 1987.

Cramer G, "Automonous Electrical Power Supply Systems", Proceedings of Workshop on Wind/Diesel Systems, June 1987. Published by BWEA.

Coonick AH, Bleijs JAM, Infield DG, "Wind/Diesel System with Flywheel Energy Storage", Proceedings of Workshop on Wind/Diesel Systems, June 1987. Published by BWEA.

de Bonte JAN, Costa JL, "An Autonomous Wind Diesel System on the Cape Verdian Islands: Design, Testing and Practical Experiences", Proceedings of Workshop on Wind/Diesel Systems, June 1987. Published by BWEA.

Van de Hoven I (1957), "Power Spectrum of Horizontal Wind Speed in the frequency range from 0.0007 to 900 cycles per hour" Journal of Meteorology, Vol 14, 160-164.

Slack GW, Musgrove PJ, "A Wind Diesel System with Hydraulic Accumulator Energy Buffer", Proceedings of Workshop on Wind/Diesel Systems, June 1987. Published by BWEA.

Bullock AM, Musgrove PJ, "Computer Modelled Performance Characteristics of a 60kW Wind/Diesel System", Proceedings of Workshop on Wind/Diesel Systems, June 1987. Published by BWEA.

Somerville WM and Stevenson WG, (1984), "Optimal use of Wind and Diesel Operation on a Remote Scottish Island" in Proceedings of the EWEC 1984 European Wind Energy Association Conference (ed W Palz) Oxford, Cotswold Press.

Somerville WM and Stevenson WG (1986) "An Independent Wind Powered Generation System with Pumped Storage and Diesel back-up" Proceedings of the EWEC 1986 European Wind Energy Association Conference, Rome, (ed W Palz).

Linders J, Holmblad L, Andersson B, "Current Progress with the Autonomous Wind-Diesel System at Chalmers University", Proceedings of Workshop on Wind/Diesel Systems, June 1987, Published by BWEA.

Schott T, Zeidler A, Reiniger K, "Hybrid System Wind/Photovoltaic/Diesel/ Battery - Theoretical and Experimental Results", Proceedings of Workshop on Wind/Diesel Systems, June 1987, Published by BWEA.

Lipman NH, De Bonte JAN, Lundsager P, "An Overview of Wind/Diesel Integration: Operating Strategies and Economic Prospects", European Wind Energy Association Conference and Exhibition 7-9 October 1986, Rome, Italy.

Halliday JA, Gardner P, "Wind Integration for Large Multiple-Diesel Island Systems", Proceedings of workshop on wind/diesel systems, June 1987, published by BWEA.

Lundsager P, "Implementation of a 55/40kW Wind/Diesel System with Energy Storage in Cape Verde, Proceedings of Workshop on Wind/Diesel Systems, June 1987, Published by BWEA.

Infield DG, and Puddy J, "Wind Powered Electricity on Lundy Island, Energy for Rural and Island Communities III, published by Pergammon Press, Oxford.

Twidell JW, Anderson GA, Gardner P, Halliday JA, Holding NL, Lipman NH (1987), "Wind Generated Power for Shetland: Tactical planning for the 30MW Peak Autonomous Grid and Diesel/Thermal Plant". Proceedings of the 9th BWEA Conference, held at Edinburgh. Published by Mechanical Engineering Publications, Bury St Edmunds, Suffolk.

Contaxis GC, Kabouris J, Chadjivassiliadis J, "Optimum Operation of an Autonomous Energy System" European Wind Energy Association Conference and Exhibition, 7-9 October 1986, Rome, Italy.

Bass JH and Twidell JW, "Wind/Diesel Power Generation - Strategies for Economic Systems", Proceedings of 8th BWEA Annual Conference (ed Anderson and Powles), 1986.

Dure JD, "System Design for Diesel/Renewable Hybrid Power", Mechron Energy Ltd, Ottawa, Canada, 1985.

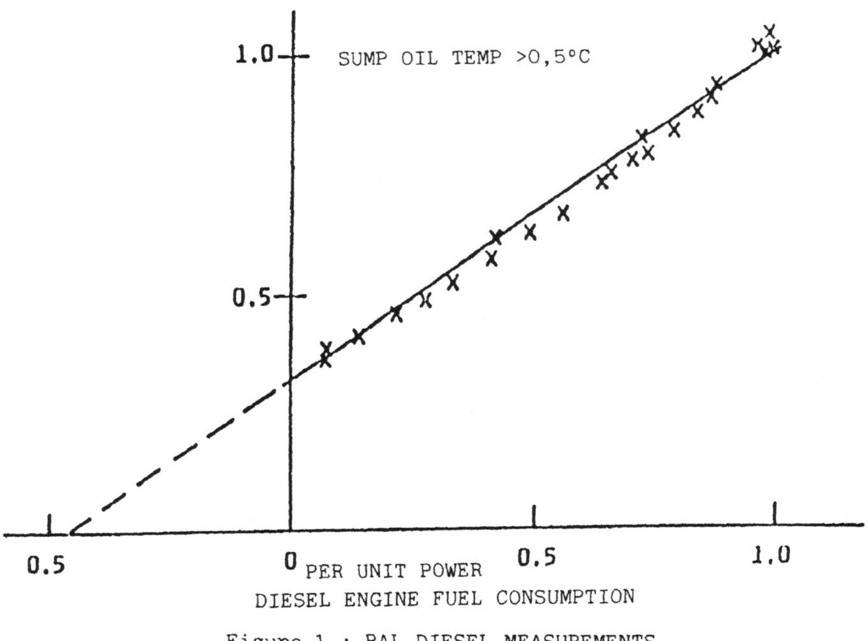

Figure 1 : RAL DIESEL MEASUREMENTS

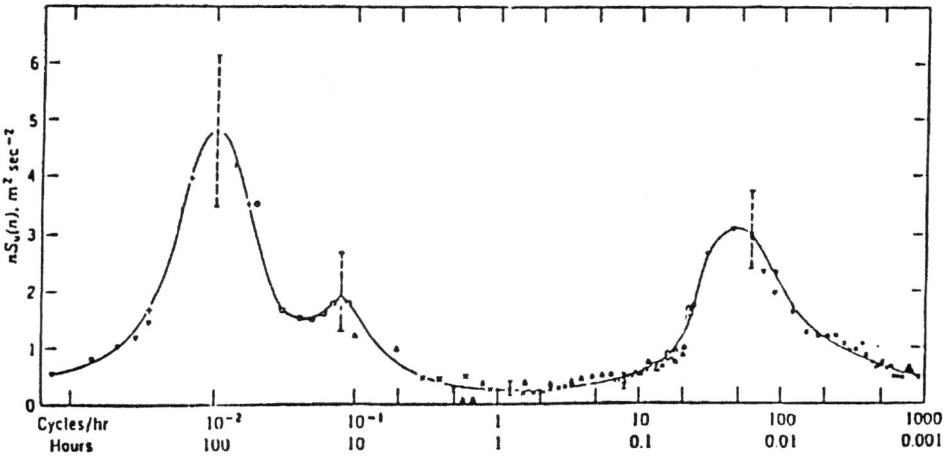

Figure 2 : Schematic spectrum of wind speed near the ground estimated from a study of Van der Hoven (1957)

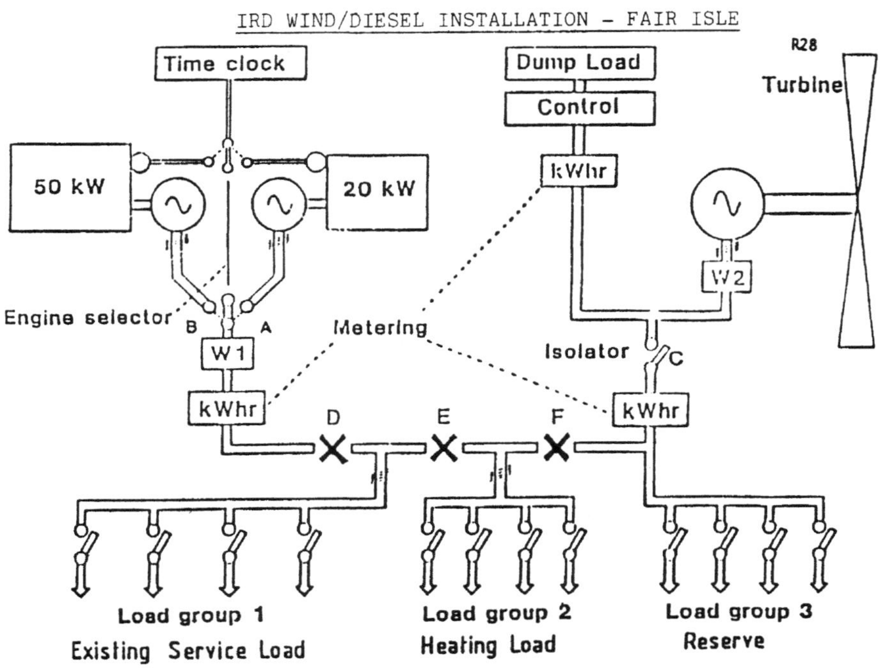

Figure 3 : Wind and Diesel Distribution arrangement

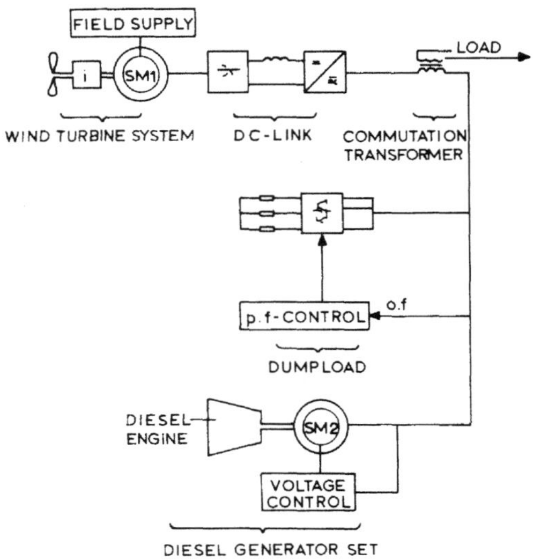

a. Outline of the AWDS

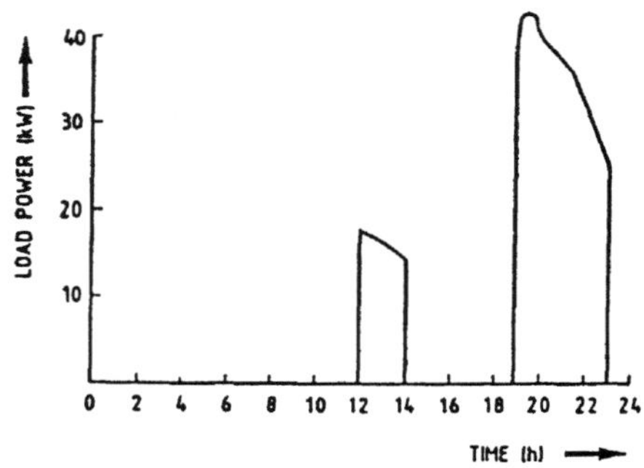

b. The load power pattern of the village of Tarrafal

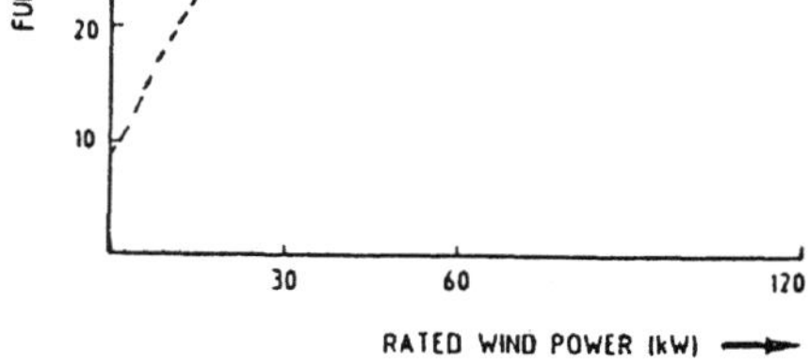

c. The expected fuel savings as a function of the installed wind power

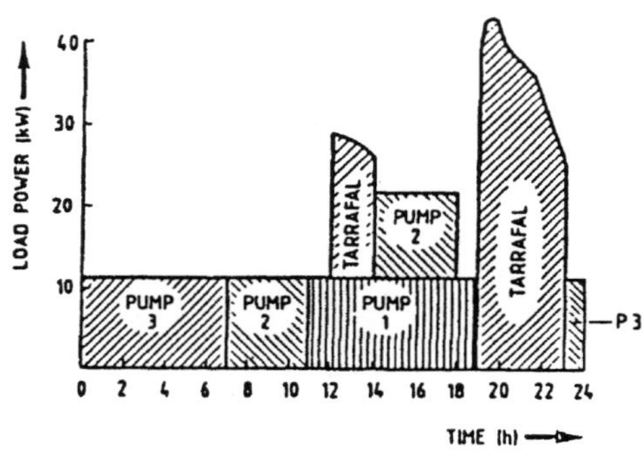

d. The load power pattern for the AWDS

Figure 4 : CWD Project in Cape Verdian Islands

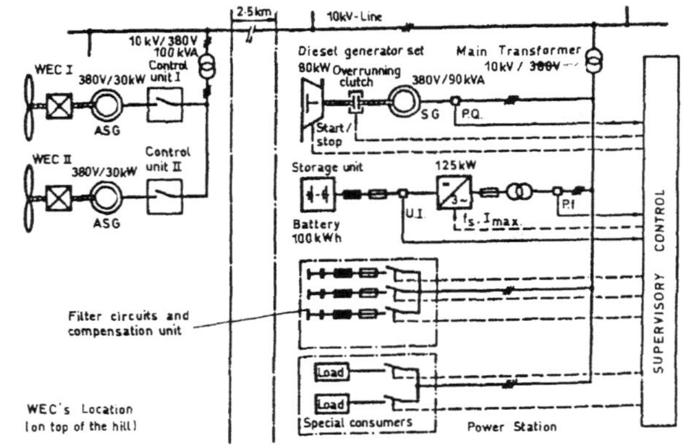

Figure 5 : Design and dimensioning of the autonomous electrical power supply system for Cape Clear Island

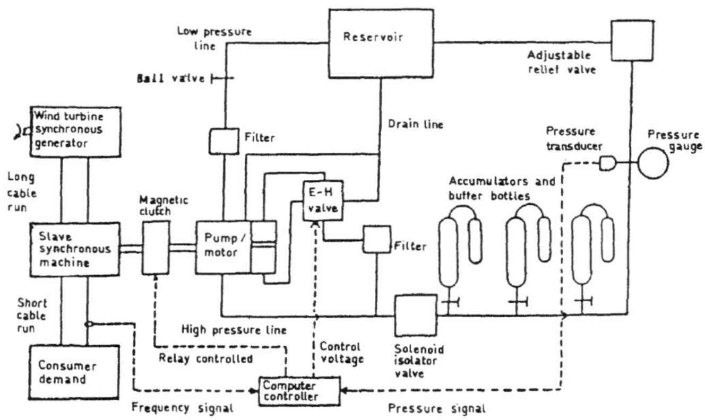

Figure 6a : Schematic diagram of the Hydraulic Accumulator System

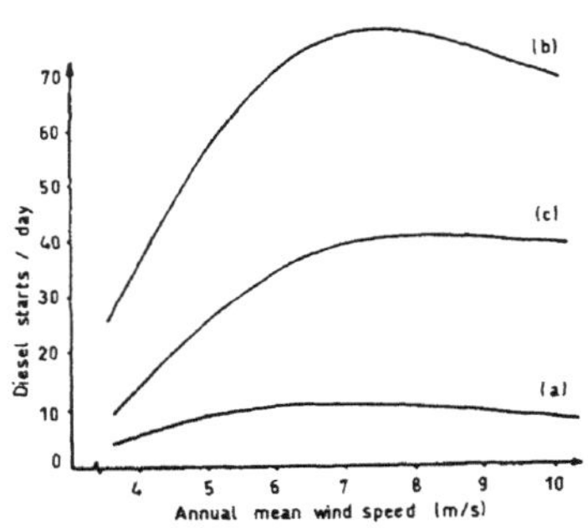

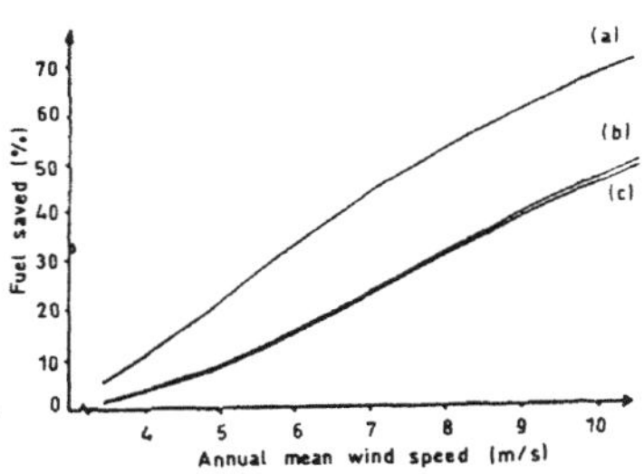

Key : a) with hydraulic store described in Fig. 2
b) No storage ; 5 minutes minimum diesel runtime
c) As (b) with 10 % hysteresis + 5 minutes averaging

Figure 6b : Diesel starts versus annual mean wind speed Rayleigh Distribution; 4 kW Wind Turbine 1 kW average load ; Load Factor 0.4

Figure 6c : Fuel saving versus annual mean wind speed Key : As previous figure

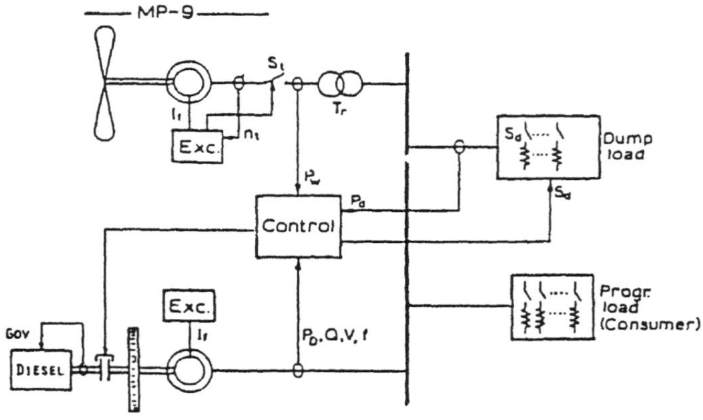

Figure 7a : Wind/diesel power system with flywheel storage

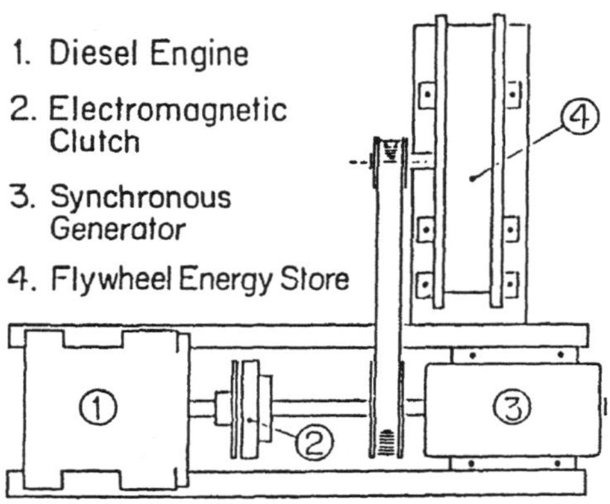

Figure 7b : Lay-Out of diesel/flywheel test rig

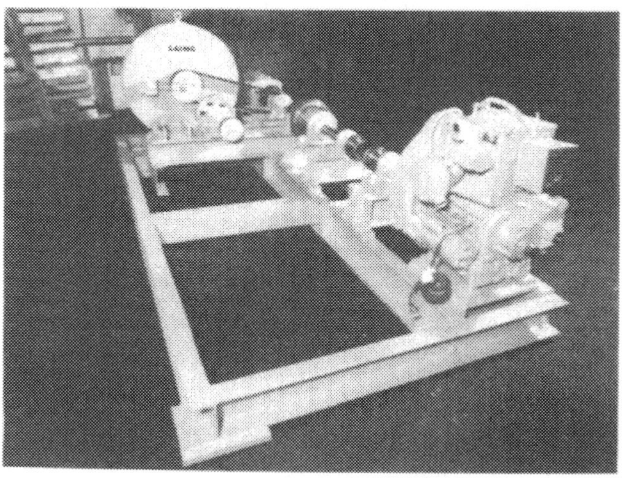

Figure 8 : Diesel-flywheel rig (under test at John Laings Ltd.)

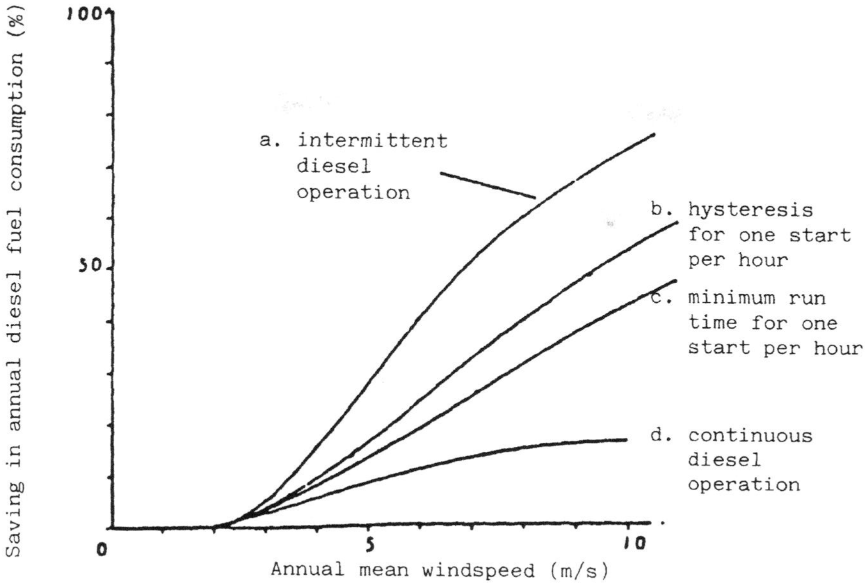

Figure 9 : Annual diesel fuel consumption versus annual mean wind speed for :

a) intermittent diesel
b) hysteresis
c) minimum run time
d) continuous diesel

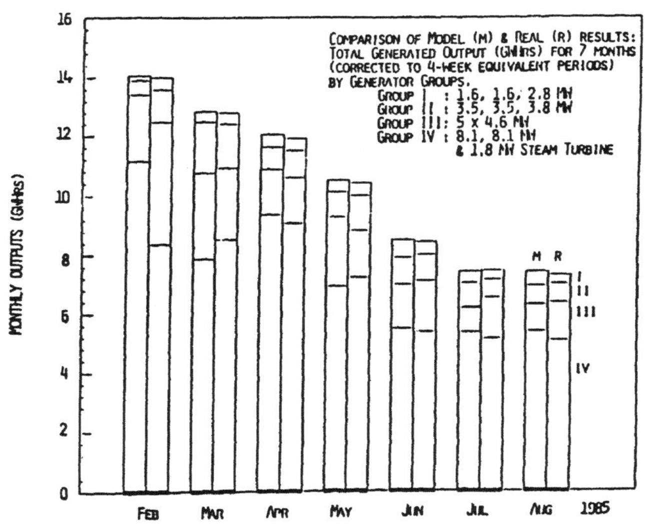

Figure 10: Comparison of model (M) & real (R) results

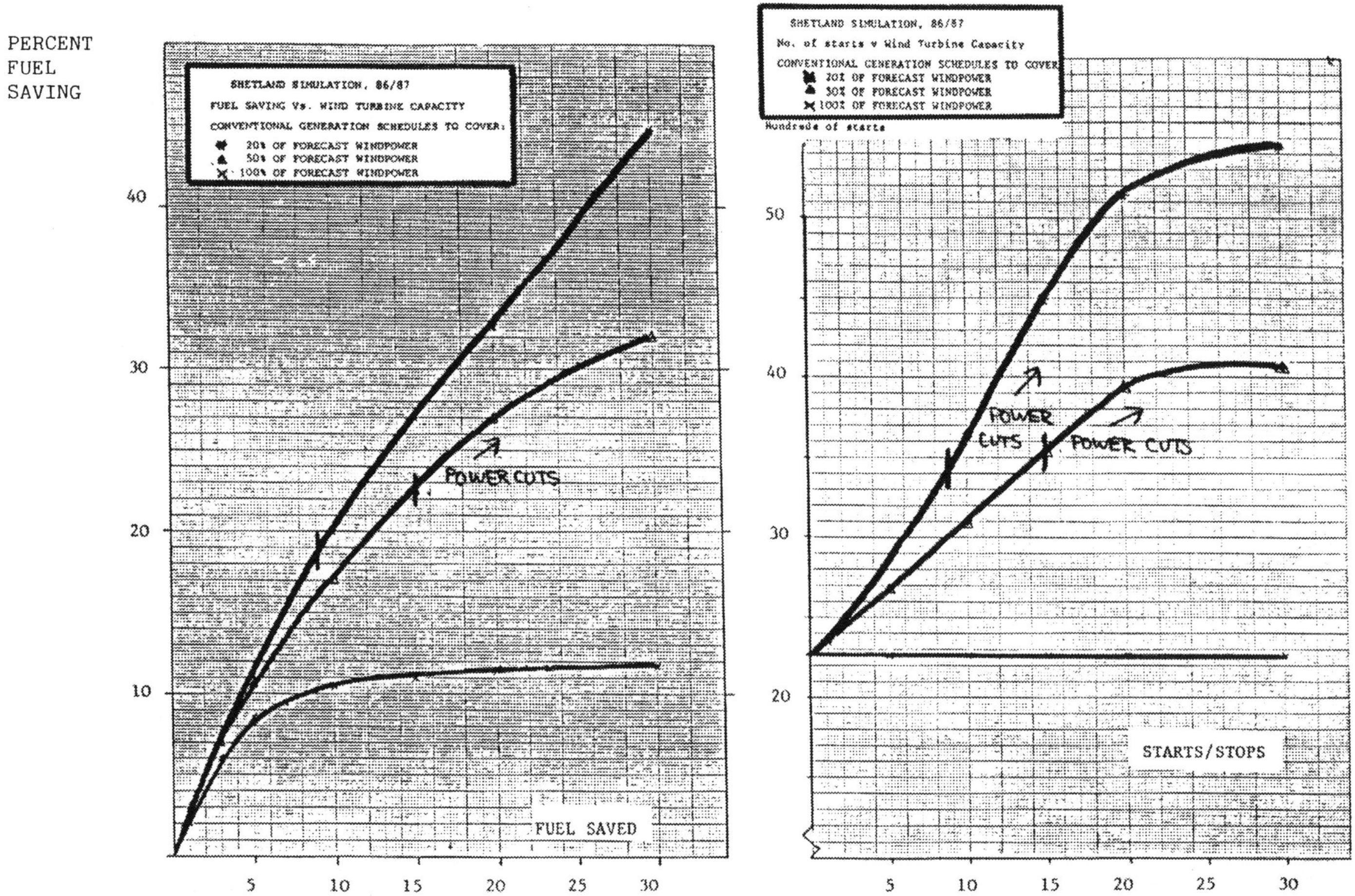

Figure 11 NUMBER OF WIND TURBINES

NUMBER OF WIND TURBINES

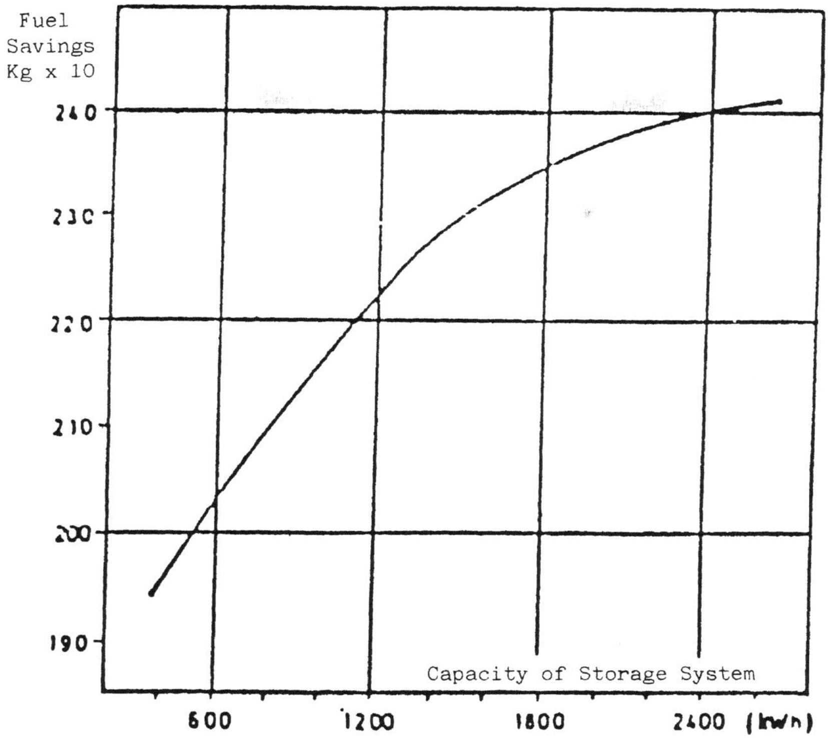

Figure 12 : Fuel Savings vs Capacity Level of Storage System

CONTROL AND LOAD MANAGEMENT SYSTEMS ON WIND POWER PLANTS CONNECTED TO DIESEL BASED GRIDS

Dipl.-Ing. G. Cramer
SMA Regelsysteme GmbH
Hannoversche Str. 3, 3501 Niestetal 1
Federal Republic of Germany

1. Introduction

Due to the high cost of generating electric power in decentral power supply systems using diesel generator units, an economical operation of small electric power supply systems using the power of wind is already possible nowadays.

The use of wind energy converters in weak isolated grids not only makes high demands regarding reliability, but also makes necessary a control and system design exceeding essentially the complexity of grid-parallel operation.

The demands on control and electrical equipment of wind energy plants in isolated operation, and the configuration of the systems are described in the following.

2. Isolated Operation of WECs

In isolated operation the wind energy plant directly supplies electricity consumers. Two fundamental plant types can be distinguished:

1. Wind energy plants with fixed rotor pitch
2. Wind energy plants with controllable blade pitch angle

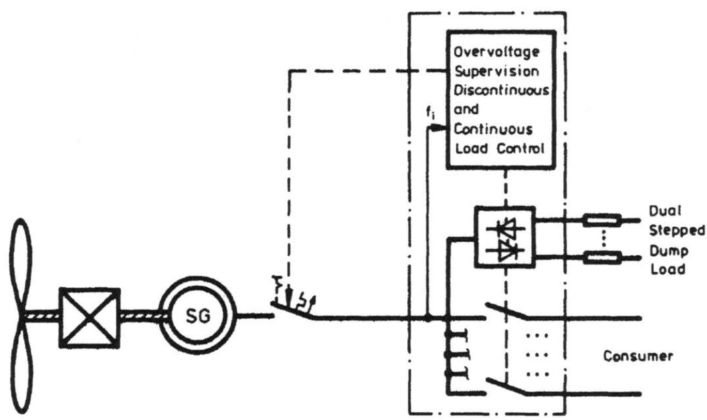

Figure 1: Arrangement for the connection / disconnection of load steps, and additional continuous control of a load step for fine speed control

2.1 Isolated operation of one WEC with fix pitch

Due to the great expenditure for a controllable blade pitch setting, small wind energy plants are often built with constant blade pitch angle.

The generator must be designed in a way that it can transform the power offered by wind up to the shut-down wind velocity v_{ab}. A mechanical shutdown device (e.g. flaps or breaks), or respective aerodynamical design brings the plant above v_{ab} to a standstill. Besides the necessary protection against overload for the generator, the consumers have to be protected against overvoltages during overspeed of the plant by disconnection.

Figure 1 shows the arrangement for a wind energy plant with constant blade pitch angle, with generator frequency control by means of a fast connection or disconnection of additional dump loads.

The connection/disconnection of load circuits is carried out priority-controlled and discontinuous by means of contactors. The priorities and the connection/disconnection criteria are set by means of the small microprocessor system keyboard separately for each load step. Finer speed control is obtained by a stepless control of the power input from other load circuits by means of dually stepped resistors. Semiconductor switches must be used here due to the greater switching frequency. Speed variations still occurring are in a range of $\pm 3\%$ for this arrangement. Power consumed by the additional load circuit may e.g. be used for heating.

2.2 Isolated operation of one WEC with pitch control

An essential advantage of plants with blade pitch setting device is, besides the better starting action and the greater efficiency, the possibility to operate it at very high wind velocities independent from the actual consumer power.

Precondition for a speed control by a variation of blade pitch angle is always that the power offered by wind is greater than the sum of converter losses and consumer power to be supplied.

That's why two operating ranges are to be distinguished in isolated operation of wind energy plants:

- Nominal load range
  Operation with wind velocities above rated value, in which the plant is speed-controlled by a variation of blade pitch angle.
- Partial load rang
  Operation with wind velocities below rated value; the maximum power supplied by the wind is smaller than nominal power.

The power to be supplied must always be kept lower than the power offered by wind, if speed is to be kept constant by the variation of the blade pitch angle even in the partial load range. For this purpose the consumers to be supplied are distributed as proportionate as possible to a number of different load circuits, which are connected/ disconnected depending on frequency and frequency gradient. Figure 2 shows this kind of wind energy plant with electro-hydraulic speed control and consumer control.

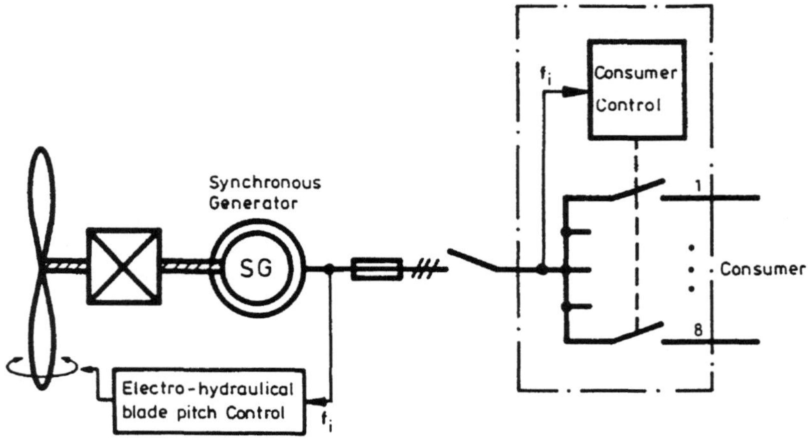

Figure 2: Control method on principle for wind energy plants with controllable blade pitch angle in isolated operation

3. Wind park in isolated operation

Besides an increase of availability of the compound system, the installation of wind parks has the great advantage, that the short-time variations of output power due to wind energy offer variations (in the seconds range) are better balanced.

The parallel operation of several wind energy converters for the supply of greater consumer power should, if possible, be carried out with wind energy converters having a fast speed and power control by a variation of blade pitch angle and an induction generator.

Here the installation of wind energy converters with fixed pitch is also possible on principle, if power limitation and speed control via dump loads is intended. In this case a dump load speed control may be used commonly by several converters.

The design of all wind energy plants as grid connected plants with asynchronous generators is useful. A commonly used rotating phase-shifter, and a connectable reactive power compensation unit, which is e.g. installed in a control room, take over voltage control and supply of reactive power to the consumers.

The wind park is designed for two wind energy converters. A third converter may be installed later if required. The wind energy converters are standard-type versions for grid operation with asynchronous generators. The rated power of each converter is 30 kW.

Each plant has a speed and power control by means of a fast-working pitch control. Voltage control is carried out by a rotating phase-shifter with a rated power of 40 kVA, which stands on the floor. For starting after dead calm periods, the phase-shifter is driven up to nominal speed by means of a small DC motor. In this way it can build the grid conditions necessary for the wind energy converters, so that they can start up. If one of the plants has connected its generator to the grid, the DC motor is switched off and mechanically disconnected by means of an overrunning clutch.

A battery, buffered by a charger during the operation of the wind energy plants, supplies the DC motor. To avoid unnecessarily frequent starting, and unnecessarily great battery capacity, a starting attempt is carried out only if the measured wind speed is sufficient.

To unload the rotating phase-shifter, an additional controllable compensation device is intended.

The commands for the connection of the several pumps and for starting the rotating phase-shifter are given by a consumer control. The logic is based on a compact microprocessor system, so that all switching criteria may be altered even at the plant's location.

The system is shown in figure 3.

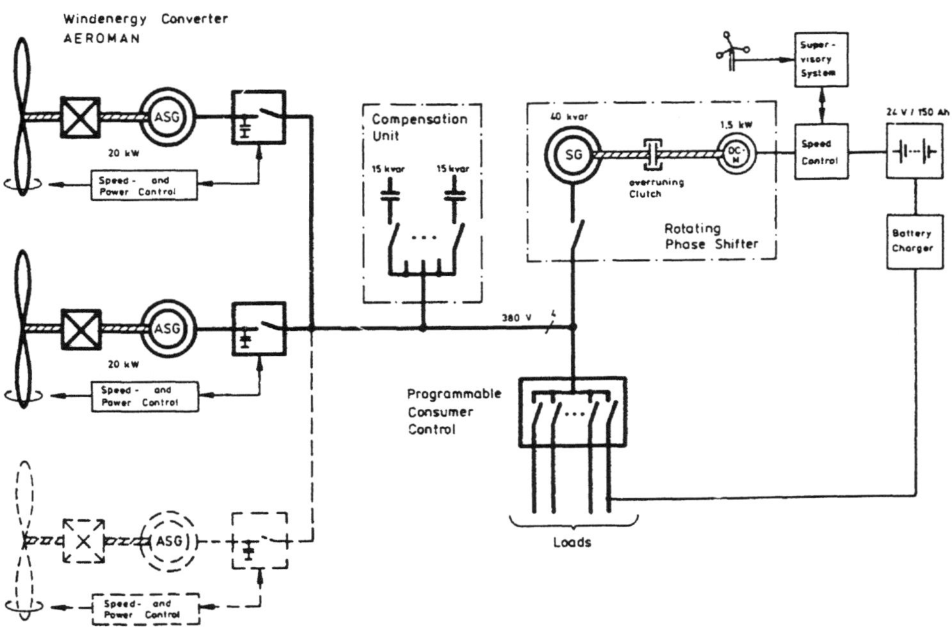

Figure 3: Block diagram of a wind park in isolated operation (e.g. for water pumping)

4. Wind/Photovoltaic/Battery-Combination

The enlargement of the wind park described before by a solar generator and a battery storage is useful, if a supply of essential consumers shall be maintained even during dead calm periods.

The supply system consists of 2 wind energy converters of the AEROMAN-type with a nominal power of 25 kW each, a photovoltaic generator with a nominal power of 15 kW_p, and a battery system with a capacity of about 280 kWh, which is coupled by means of a converter with the 3-phase AC-bus (v. fig. 4). In the following the components are described in detail.

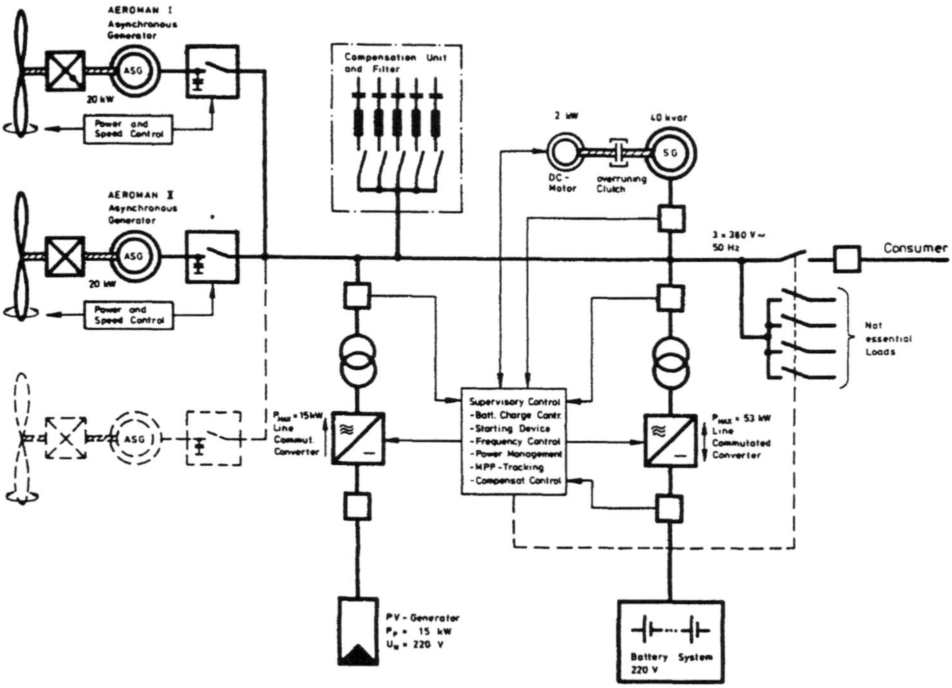

Figure 4: Block diagram of the Wind/Photovoltaic/Battery-Combination for a village supply system

Wind Energy Converters

The wind energy converters AEROMAN with a rotor diameter of 15 m are speed and power controlled by a fast rotor blade adjusting device, and equipped with an asynchronous generator for grid operation in the standard version.

The wind energy converters are directly connected with the 3-phase AC-bus, to which also the several loads of the village supply are connected. In this way it is prevented to lead the power supplied by the wind energy plants through converter, battery system, and inverter. Far better efficiency is reached so.

The intended wind energy plants may be connected to every location of the 3-phase AC grid you like. They don't need additional control lines, and connect themselves automatically to the grid formed by converter and rotating phase-shifter depending on wind velocity.

Photovoltaic Generator

A unit with 15 kW_p at 220 V output voltage is intended as photovoltaic generator. Its power is supplied to the isolated grid by means of a line commutated converter with transformer. A maximum power point control guarantees optimum energy yield.

Battery and Converter

The lead-battery has a capacity of 1500 Ah at a nominal voltage of 220 V. To maintain long battery life, a respective control unit is intended. This battery control acquisits the battery currents and the battery voltage, and calculates the batteries' charging state from these values. Thus an optimum working of the battery is made possible; deep discharging and overcharging are prevented.

A line commutated inverter is used as converter, which stands out for robustness and low price. This converter takes over frequency control for the 3-phase AC grid either by feeding power from the battery to the grid, or by charging the battery with surplus energy from wind plants or PV-generator.

Rotating Phase-Shifter

Voltage control is taken over by the rotating phase-shifter, which is started by a small auxiliary DC motor. After starting, this motor is switched off and mechanically decoupled by an overrunning clutch. The power for the starting process comes from the battery.

Another advantage of this conception is the possibility to make usual grid fuses break by means of the rotating phase-shifter. If faults occur in the 3-phase AC grid, the rotating phase-shifter can feed overcurrents for a short time, which make usual fuses break. This would not be possible with static converters, because these are current-limited and feed with nominal current to the failure for a longer time so that danger for man or plant cannot be avoided.

A switchable compensation unit reduces the reactive current loading of the phase-shifter. The steps of the compensation device are designed as draining circuit filter so that a reduction of the harmonic content caused by the converter takes place at the same time.

Control and Supervision

Control and supervision are carried out by a compact microprocessor system. It is responsible for an optimum and safe operation of all components. Therefore the following tasks have to be taken over by this system.

- battery control
- starting of the rotating phase-shifter
- frequency control
- power management
- connection and disconnection of the compensation steps
- MPP-control for the PV-generator

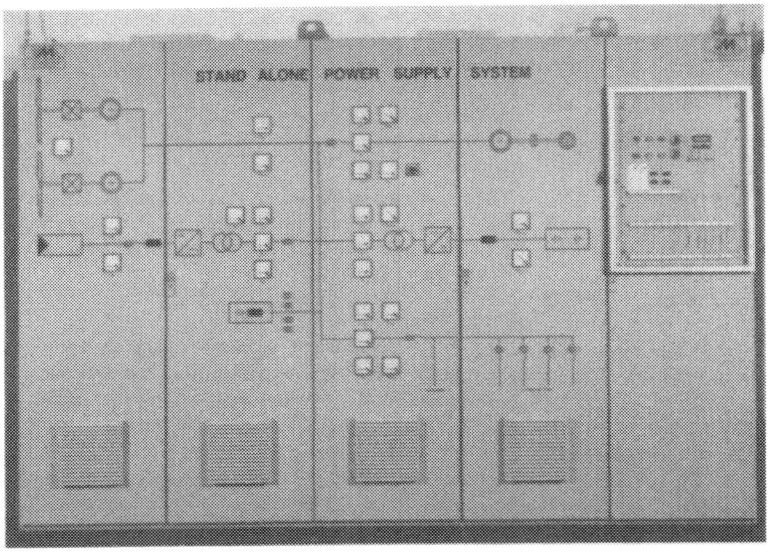

Figure 5: Switching cabinet for the Wind/Photovoltaic/Battery Combination

5. Wind/Diesel-Systems

If a safe supply shall be maintained, a diesel engine and a battery storage are necessary for the supply during dead calm periods. The use of a battery for long-time storage is not economical, that's why the battery system in the following wind-diesel-conception is installed only as short-time storage.

During the parallel operation of one or more wind energy plants with a weak electrical supply grid, which is supplied by one or more small diesel-generator units, a multitude of special requirements occurs concerning the control and supervision of the several wind energy plants, which exceeds the requirements of the parallel operation with the electrical supply grid by far.

In particular, a fast working speed control of the wind energy plant, a smooth grid connection and good power control are necessary. This makes possible an optimized smoothing of the electrical power output as well as a continuous limitation of output power in special cases of operation.

The speed/power control must be carried out by means of continuously controllable dump loads again when using wind energy converters without pitch control.

To avoid an inconvenient unloading of the diesel engine (e.g. operation below 25 % of the rated power), the power output of the wind energy plants must be reduced during light-load periods or good wind conditions. The grid frequency serves as an indicator for the loading state of the diesel engine, because the frequency of the electrical power supplied by the diesel decreases with increasing loading due to the control action of the speed governor. In this way the power output of the wind energy converter can be reduced continuously above a specified frequency value, and an inacceptable unloading of the diesel can be prevented.

5.1 Fuel Saver Operation

The conception described in this chapter is the most simple combination of a wind energy plant with a diesel engine (see Fig. 6). The wind energy plant is always operated parallel with the diesel engine. It supervises the grid conditions automatically and switches itself on/off depending on the wind speed. The WEC works as a normal grid connected machine, and frequency - and voltage control is taken over by the Diesel-Generator set.

Two load circuits of lower priority at least should be disconnectable to make possible a simple load adaption.

The limitation of wind energy converter power depending on diesel power, i.e. depending on the frequency must also be possible for this quite simple control to avoid inadmissible unloading of the diesel engine.

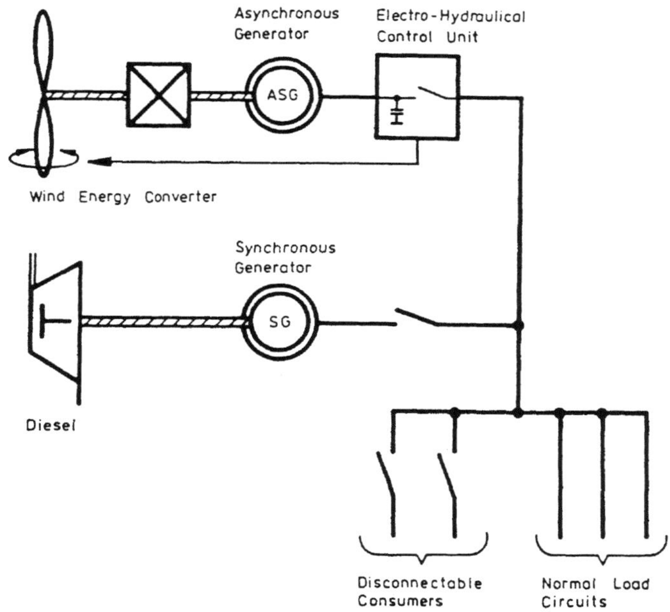

Figure 6: Block diagram of fuel saver operation

Advantages
- Simple and robust conception
- No additional electronic control units
- Easily extendable

Disadvantages
- Operation of the wind energy plant possible when the diesel is on
- Uneconomical partial-load operation of the diesel engines cannot be avoided

5.2 Systems with Battery Storage and Disconnectable Diesel Engine

This design makes possible a supply even without running diesel engine during good wind conditions or during low-load periods. The diesel engine is equipped with an electrically controllable clutch or a mechanical overrunning clutch instead of a fixed coupling of motor and generator, so that the synchronous generator works as a rotating phase-shifter when the diesel is off, and takes over voltage control as well as the supply of reactive power for the pre-compensated asynchronous generator of the wind energy plant (v. Fig. 7). During weak-load periods and good wind energy supply it is possible hereby to shut the diesel down completely and to carry out supply by means of the wind energy plant alone.

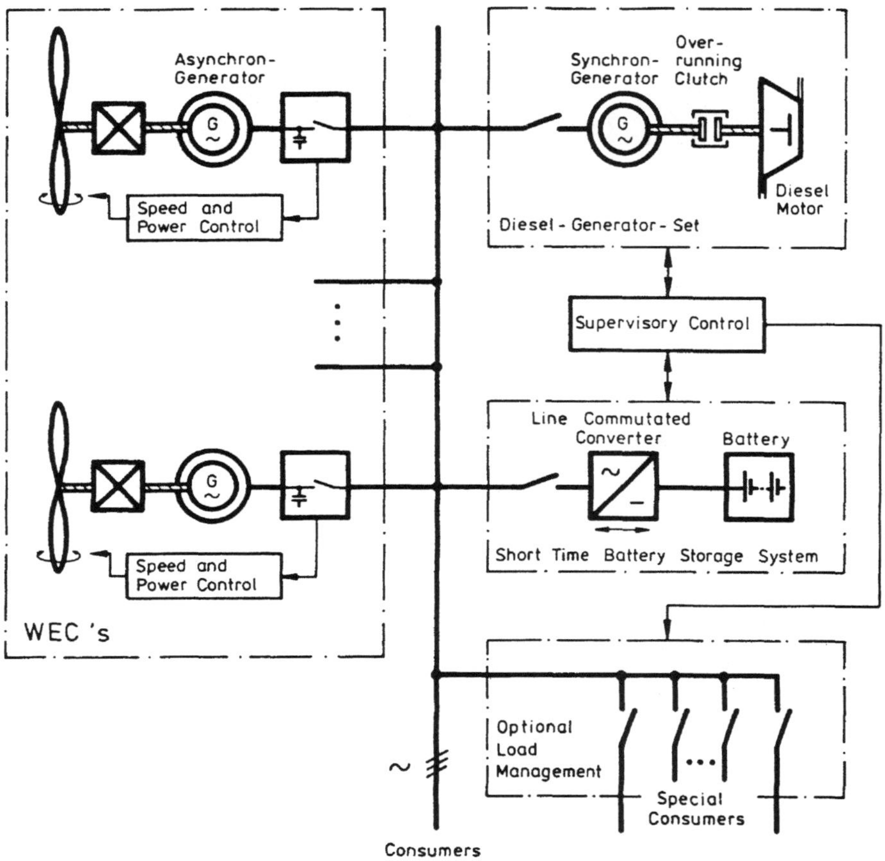

Figure 7: Simplified block diagram of the modular system for an autonomous electrical power supply (with static converter)

To avoid an unnecessary frequent starting of the diesel engine caused by changes of wind-speed or consumption, the system is equipped with a battery storage. The capacity of this storage is rather small (appr. rated power for about 30 minutes), and depends on the local conditions and the consumer's demands. For the charging of batteries and to supply the isolated grid, a line (machine)-commutated converter is installed, which stands out for simple construction and low price. Now the installation of the battery storage makes possible an optimized use of the energy supplied by the wind energy plants and a renunciaton of a fine graded consumer control. Only one or two controllable load circuits dependent on wind and consumer conditions may be installed, to increase the economy of the system during low load time.

Furthermore, the installation of a storage unit makes possible operation of the diesel engine in more suitable power ranges for a longer time, because e.g. load peaks, which would make necessary starting of the diesel, can be checked by the storage unit. Because load peaks exceeding the rated power of the diesel normally occur only for short times, the diesel engine may be dimensioned smaller. The load peaks are checked by the battery storage then.

Operation Modes and Control Methods

The four essential operation modes differ in the mode of frequency control, while voltage control is maintained by the synchronous generator, which is always working. The plant is started up with the diesel engine. The different modes are:

Parallel operation of WECs, diesel engine, and battery storage
During periods with low wind velocity and high consumer power demand.

Parallel operation of diesel engine and battery storage
During periods with unsuitable wind conditions.

Parallel operation of WECs and battery storage
During periods with sufficient wind velocity.

Single operation of the battery storage
During short periods with low wind velocity and low power consumption.

The control of the power supply relation between different components of the system works on the basis of a frequency / power static, so that the actual grid frequency is taken as base value for the plant's supervisory unit.

The most effective operation mode regarding economy and supply safety is chosen by the microprocessor-equipped operation control system depending on the actual power output of the plant, the battery's charging state, the actual load conditions, and the expected consumer power.

5.3 Systems with several Diesel Power Sets

It is useful to install several diesel power sets with different power, if a consumer power in the range from 0.1 to 2.0 MW is to be supplied. A series of optimization possibilities results by selecting the diesel power set most suitable for the supply of the actual consumer power, especially for systems with short-time battery storage. Main criterion for a minimization of fuel consumption is to avoid the operation in the range of idling or partial-load losses. Variations of consumer power and wind energy offer complicate the determination of the most suitable battery combination during operation. However, the necessary diesel reserve can be kept very small by using the battery as spinning reserve.

Figure 8 shows the characteristic fuel consumption curve of a diesel power station with 6 sets depending on the electric loading. The fuel consumption consists of an idling part and a power-dependent part [2].

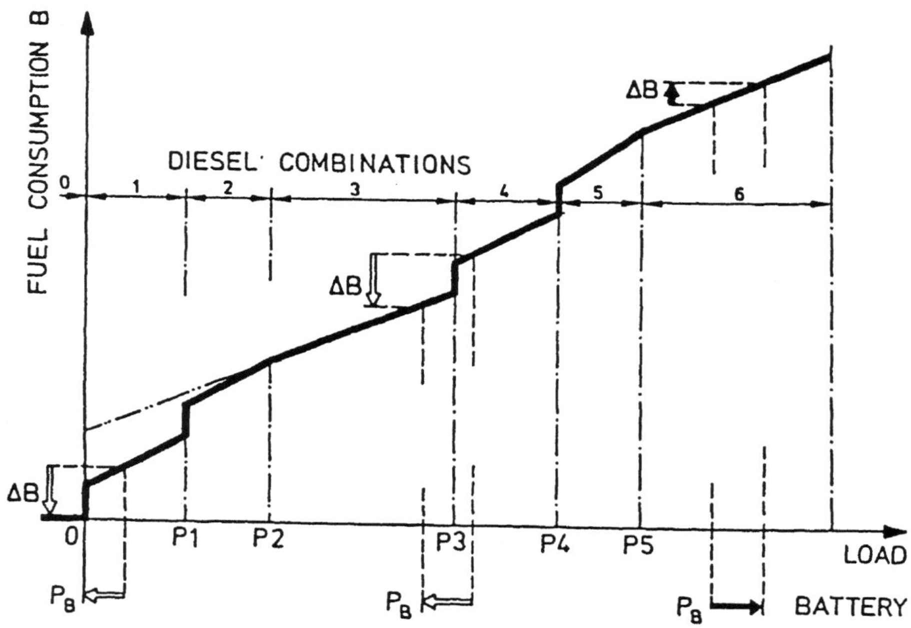

Figure 8: Fuel consumption of the diesel power set as a function of loading

6. Investigations

A series of theoretical and practical preliminary investigations regarding the design of the autonomous power supply system was carried out by SMA-Regelsystme GmbH in co-operation with the department of electrical engineering at the University of Kassel (Prof. Dr. Ing. W. Kleinkauf). Investigations into the influence of dynamical loading of diesel engines had the result that the increase of fuel consumption during parallel operation of wind energy converters and diesel engines can be neglected. The part-time shutdown of the diesel engines causes only an exceedingly small surplus consumption during the warming-up phase.

The simulation program implemented in the process computer of the department makes possible the investigation of dynamic actions by means of a simulation model of the machine commutated static converter unit.

By means of function tests with a wind-diesel-battery storage system in the laboratory of the University Kassel experience regarding frequency/voltage control was gained. Measures for the reduction of voltage distortions caused by the static converter, and for the minimization of losses in the synchronous machine due to the reactive power loading were tested.

References

[1] G. Cramer, R. Grebe, B. Hanna, J. Sachau
Advanced Autonomous Electrical Power Supply for the small Irish Island of Cape Clear
EWEC'86, European Wind Energy Association, Conference and Exhibition
7-9 October 1986, Rome-Italy

[2] J. Chadjivassiliadis, G. Hackenberg, W. Kleinkauf, F. Raptis
Power Management for the Compound Operation of Diesel Generator Sets with Energy and Photovoltaic Plants
EWEC'86, European Wind Energy Association, Conference and Exhibition
7-9 October 1986, Rome-Italy

[3] W. Kleinkauf (et. al.)
Final Report for BMFT-Project
"Combined Operation of Wind Energy Converters"

Project Nr 127/83 UK

THE DEMONSTRATION OF A 100KW VERTICAL AXIS WIND TURBINE

I.D. Mays, C.A. Morgan and M.B. Anderson
Vertical Axis Wind Turbines Limited

Summary

This paper describes the construction, erection and demonstration of an innovative 17m diameter vertical axis wind turbine on the Isles of Scilly, off the south west coast of the U.K. mainland. The wind turbine has a rated output of 100kW and is intended for commercial production.
Electricity on the Isles of Scilly is currently generated by diesel sets although a connection to the mainland is now being installed. With the substantial variation in load upon the small grid network on the islands, experience may be gained both at low penetration levels during the day and medium to high penetration levels at night.
The wind turbine was first erected on 20th June 1987 and experience gained during the summer and autumn. Overtravel of the reefing mechanism in October has led to damage of the rotor. As a decision has now been made by VAWT Ltd to build future machines with fixed geometry and stall regulation, the rotor for the Scilly Isles wind turbine is being modified to this configuration and the unit will be reinstated in the summer of 1988.

1. INTRODUCTION

The development of straight bladed vertical axis wind turbines commenced in the U.K. in 1975 with early experimental work being undertaken at Reading University using a prototype 3m in diameter. In 1978 the potential of this type of wind turbine for scaling up to larger sizes was recognised and the U.K. Department of Energy provided support funding for a phased programme of development. The aim of the programme was to provide technology which would enable the construction of multimegawatt size units offshore for centralised generation of electricity. Initial work was undertaken by an industrial consortium which was subsequently formed into a company, Vertical Axis Wind Turbines Ltd. This company is jointly owned by Sir Robert McAlpine & Sons Ltd and Northern Engineering Industries plc. The first major phase of the programme was to design and construct a 25m diameter research prototype machine to be used to refine design parameters and gain experience in advance of the construction of larger machines. Construction of this machine was completed in September 1986 and 18 months operating experience has now been gained. The wind turbine has proved to be extremely reliable with availability during 1987 being 94%.

In 1983 VAWT Ltd recognised that a substantial market was developing worldwide for medium sized wind turbines both for grid connected use and operating in parallel with diesel engines. It was decided therefore to utilise the innovative technology being developed for large straight bladed vertical axis wind turbines to demonstrate a commercial derivative rated at 100kW which could subsequently be marketed worldwide. As with the 25m research prototype the proposed machine had a variable geometry rotor in order to provide power control and to minimise structural loading.

A significant portion of the future market was foreseen as associated with operation into diesel fired networks on islands and other remote communities. It was therefore proposed that the demonstration machine be sited on an island with a small diesel powered grid. Sir Robert McAlpine & Sons Ltd were closely involved with the assessment of energy usage on the Isles of Scilly undertaken in the early 1980's. It was therefore appropriate that the demonstration machine be sited on these islands where it would be possible to gain experience with the integration into the diesel fired network both at high, medium and low levels of penetration. After agreement with the local authorities the selected site was chosen on the north east corner of the largest island, St Marys at Mount Todden (Figure 1).

2. ST MARYS

The Scilly Isles are situated 45km off the south west tip of England. Five of the islands are inhabited and have a total population of approximately 2000. The largest of these is St Marys where an electrical grid network is operated by the U.K.'s South Western Electricity Board. Of the other islands only Tresco has a small grid network which is privately operated. Inhabitants of the remaining island utilise individual diesel generating sets. Being a holiday area, the population swells substantially during the summer months resulting in a seasonal demand for electricity which varies considerably from that on the mainland.

The generating station on the island, operated by the South Western Electricity Board comprises a range of diesel generating sets which is provided to meet the wide variation in demand. During the day this varies between 1MW and 2MW, supplied from an appropriate combination of plant, but at night drops to approximately 300kW when only one 450kW generating set is operating. Having a rated power of 100kW, the wind turbine therefore provides a modest proportion of the capacity during the day but, in wind speeds above the rated, could produce one third of the demand at night. Valuable experience may therefore be gained both at low and higher levels of integration of wind power into the diesel network. Considerable care is, however, necessary to ensure that the introduction of the wind turbine does not reduce the standard of supply on the island. Mathematical models have been prepared to analyse the impact of the wind turbine upon the grid especially the effect upon voltage transients during start up and under varying operating conditions and penetrations of demand. In order to meet these requirements the wind turbine is started upon the smaller 30kW generator.

After a recent review the South Western Electricity Board decided to inter-connect the smaller islands by subsea cable to the power station on St Marys. This was undertaken during 1986 and has had the effect of increasing the demand on St Marys Power Station by 10%. As a result the capacity on St Marys has been increased from 3.7MW to 5.6MW. In order to further improve the supply to the islands, the South Western Electricity Board has decided to connect St Marys to the mainland by subsea cable. This work is currently in hand.

Wind speed measurements have been made on the wind turbine site, at Mount Todden and correlated with the nearby meteorological station. The mean wind speed on the site has been estimated to be 7m/s at a hub height of 20m.

3. THE WIND TURBINE

Conceptual design of this 100KW innovative variable geometry vertical axis wind turbine commenced at the beginning of 1984. In the ensuing period the options on rotor configuration and tower and transmission arrangement were explored. Selection of the preferred configuration was settled in the early part of 1985 at which time VAWT Ltd entered negotiations with Davidson UK Limited, regarding their possible involvement in the project. A joint venture basis was proposed and after agreement of contractual arrangements detailed design commenced in the latter part of 1985 and was completed in the summer of 1986. Construction commenced in July 1986 and was completed by June 1987.

3.1 General Assembly

The selected configuration (Figure 2) for the wind turbine was an 'H' shaped rotor. The blades, which are supported at the end of an horizontal crossarm and also by angled struts, are inclined between 15° and 60° to the vertical by the movement of a telescopic member within the crossarm. The rotor is mounted upon a tripod tower arrangement and torque generated by the rotor is transmitted to the gearbox at ground level by a rotating tube. The weight of rotor and torque tube is supported just above the gearbox i.e. close to ground level, with the tower providing lateral restraint at the hub. The systems comprising the wind turbine are described in more detail below.

3.2 The Rotor

The material selected for the blades of the wind turbine is wood/epoxy after a review of materials appropriate for wind turbines of this size. They are similar to those used on other UK wind turbines and have been produced for VAWT Ltd by Gifford Technology Ltd of Southampton. The blades comprise a 'D' section spar fabricated from veneers of Douglas Fir laminated with epoxy. This is moulded with a light trailing edge of grp and polyurethane foam. The major structural attachment to each of the four half blades (two per side of the rotor) is midway along its span and the major part of the driving torque from the blades is transmitted to the struts at this point. The thickness of the skins are therefore at their greatest at the mid point, and taper to a minimum thickness at the blade tip and at the point of attachment of the root of the blade to the telescopic part of the crossarm. The white blades have an aerofoil section of NACA 0018 which gives better structural efficiency than the 0015 used on earlier vertical axis wind turbines. Each blade weighs 106kg.

The struts have a tapered planform having their apex where they attach to the blades in order to achieve high stiffness in the direction of rotation such that the driving torque from the blades largely passes through these components. They are formed in aluminium from a simple box structure of rivetted construction having upper and lower skins and leading and trailing edge spars. The skins are curved such that an aerofoil section results when light leading and trailing edges are fitted to the box. The attachment of the strut to the blades and the crossarm is by steel lugs and glass reinforced plastic fairings are provided at the junctions to minimise parasitic drag losses.

At the outset of the project it was specified that all components should be able to fit into a standard container for transportation purposes and for this reason the crossarm is constructed in three parts, a central part including the hub and two outer sections. A flange bolted joint is provided for assembly. The main structural member of the crossarm is a simple steel box design formed principally from two brake pressed channel sections which are welded along the neutral axis. The upper and lower surfaces of the crossarm are curved to provide an aerofoil section when leading and trailing edges in aluminium alloy are fitted. The box section is reinforced towards the hub to maximise stiffness. Corrosion protection for all steel components is provided by metal spraying and polyurethane paint finish.

The crossarm contains the mechanism for inclining (reefing) blades. This comprises a telescopic section which, moves within the crossarm and the actuation system. The telescopic member, or the reefing beam, is formed from extruded aluminium sections and is located and moves within the crossarm in a fabricated track attached to the inner skin of the crossarm. The bearing surface between track and reefing beam is a proprietary polymer sliding bearing. The reefing beam has a light alloy semicircular fairing on the leading and trailing edge to minimise drag when reefing commences. At the end of the reefing beam a hammer head fabrication is provided onto which the ends of the blades are fitted using proprietary plain bearings.

The actuating mechanism for the reefing system is attached to the reefing beam. This comprises a lead screw which is driven from a standard electric motor/gearbox on the axis of the turbine with twin output shafts, one for each side of the rotor. Transducers monitor the position of the reefing system.

The angle of the blades to the vertical is 15^{0} at all wind speeds up to the rated as this provides very similar performance to having the blades fully upright. As wind speed increases beyond the rated, 13m/sec, the blades will progressively incline to a maximum of 60^{0} when the wind speed is a minute by minute average of 23m/sec, gusting to 30m/sec. Beyond this point the wind turbine would be shut down. The system provides power regulation above rated wind speed.

3.3 The Mechanical Transmission System

The major part of the mechanical transmission system is sited at the base of the tower and torque is transmitted from the rotor via the steel torque tube. The main lateral bearings for the rotor are situated at the top of the tower. The torque tube has a stub shaft fitted to its upper end which carries the main bearings between transmission and tower. The main bearing diameter is 530mm. The torque tube is formed in two parts for transportation purposes and has an overall length of 15m. At its lower end the torque tube is attached to the lower stub shaft which passes into the main thrust bearing carrying the weight of the rotor and torque tube This is then attached to a gear coupling which is also attached to the input side of the three stage gearbox. The brake disc is mounted between the torque tube and stub shaft. The lower transmission components are mounted in a steel frame with removable panels to prevent unauthorised access. The vertically mounted gearbox has two output shafts to serve two generating systems. The braking system comprises three brake calipers acting on a single disc. Two of the calipers are used as the normal operating brake for the machine and are air operated. The third is a spring applied caliper which operates in conjunction with the normal braking system for emergency shutdown and is also used as a parking brake for the wind turbine.

Flexible couplings are provided between the output shaft of the gearbox and each generator.

3.4 The Generating System

In order to maximise energy capture from the wind turbine, two generators are provided. In low wind speeds the smaller, 30KW generator can be used. This turns at 1000rpm and corresponds to a rotor rpm of 33.3. In higher wind speeds, above 8m/sec, the larger generator operating at 1500rpm would be brought on line. This unit has a capacity of 100KW. Both generators are of the induction type having a slip of 3% to maximise the excellent power smoothing ability of the high inertia rotor. A free standing switchgear enclosure is provided adjacent to the wind turbine containing all electrical protection equipment, contactors for auxillaries and the electronic control and monitoring system.

The subsoil on the Scilly Isles is decomposed granite and it was found necessary to provide an extensive network of copper tape in the ground in order to obtain a satisfactory earth. The earthing system is used both for the switchgear and transformer and also for the lightning protection system which has been installed for both the wind turbine and the anemometer mast.

The nearest connection to the electrical network on the Island was approximately 200m away from the site and is at 11kV. An underground connection has been provided at the request of the Planning Authority and a transformer substation is situated adjacent to the machine providing a 11kV to 415V step down for connection to the generators.

3.5 Control and Monitoring System

Based upon experience gained with the 25m diameter wind turbine at Carmarthen Bay, a supervisory electronic control system has been developed as a single board to carry out the following functions:-

a) Starting of the wind turbine
b) Operation of the reefing system
c) Monitoring of the essential parameters
d) Normal and emergency shut down.

Associated with this, a single board electronic monitoring system has also been provided which allows acquisition of data from general transducers such as blade strain gauges, accelerometers and wind speeds for later analysis. Both units have been engineered by CAP Industry Limited, Reading, for VAWT Ltd.

An anemometer tower is provided at a distance of approximately 60 metres from the wind turbine to enable correlation of data taken with wind speed and also for the control of the wind turbine.

3.6 Tower

Many options on tower design and fabrication were considered to minimise the weight and cost of this component, whilst also maximising the ease of transport, particularly for island situations. Initially a guyed arrangement was considered but detailed dynamic analysis showed that whilst this was feasible, the cost would be as high or higher than a cantilever tower arrangement. The solution which was selected minimises weight whilst maximising stiffness and minimising transport problems. This is a steel tubular tripod arrangement with each leg of the tripod being formed by an 'A' frame. Each 'A' frame is split close to the centre by a flange joint such that these components may fit inside a container and are relatively easy to handle upon site.

3.7 Civil Works

The civil works on site comprise the provision of an access road, site clearance and fencing, the excavation for foundation and the provision of the foundation itself. The foundation is a gravity base comprising a reinforced concrete slab 9m in diameter and 1m thick. This was installed during April 1987.

Although trial pits had been dug early in the programme, the excavation of the decomposed granite on the site proved more onerous than expected as in places substantial pieces of rock were encountered requiring extensive blasting.

4. CONSTRUCTION

After fabrication of all components each system underwent trial assembly at the factory. The rotor was fully assembled in the vertical plane at Davidson's Works in Northern Ireland. After assembly, the reefing mechanism was exercised to ensure its correct functioning. The tower was assembled in a horizontal plane and alignment checked. The other major sub-assembly erected in the works was the mechanical transmission where the complete base frame, gearbox couplings, slip ring and generators were all assembled. The switchgear was assembled at the supplier's works. No major difficulties were encountered during this trial assembly period. The rotor was subseqently disassembled for final painting and shipped to the Scilly Isles.

The first items to arrive on the site, at the beginning of June, were the sections of the tripod tower. Whilst it is possible to erect the wind turbine using an A frame should cranage facilities not be economically available, it proved cost effective for the Isles of Scilly to use a crane. However, as their were no cranage facilities on St Marys at that time it was necessary to transport a suitable crane from the mainland with a capacity suitable to lift the rotor in one piece onto the tower top. With no roll-on, roll-off ferry this crane had to be shipped to the island by landing craft which caused some delays to the start of the work on the island.

Erection of the tripod tower was completed on 9th June by which time all other components for the system had arrived by the regular shipping service. The subsequent sequence of erection was to join, by bolting, the two halves of the torque tube which was then lifted into position within the tower. The transmission assembly was then positioned underneath and connected to the torque tube. With these items located the switchgear enclosure was installed adjacent to the tower.

Concurrent with the installation and commissioning of the switchgear, the rotor was assembled at ground level in the vertical plane. The rotor was then hoisted in one piece on 22nd June onto the tower and all systems tested with the rotor static, (Figure 3). It was turned for the first time on 27th June. Whilst the rotor was being assembled the control and monitoring system was installed. This was commissioned with the rotor before the turbine was connected to the grid.

Full commissioning using the 30kW generation system was completed on 14th July. The commissioning of the 100kW generator could not take place until sustained higher winds were available which did not occur until later in the year. Operation of the wind turbine during the summer was restricted due to very low winds from the middle of July through to the beginning of October. In addition, two weeks of operation were lost in August while a fault in the control system was investigated and rectified. Nevertheless, 60 hours operation were logged in wind speeds ranging from zero to force 8 in this period with the wind turbine generating power up to a maximum of 50kW.

Wind conditions improved in October and commissioning of the 100kW, 50rpm system commenced. This was satisfactorily achieved on 28th October. Soon after, however, a simultaneous failure in both the electronic and limit switch protection systems of the reefing mechanism resulted in an overtravel which caused some damage to the rotor.

In parallel with the operation of this machine, data gathered from the 25m wind turbine at Carmarthen Bay has shown that operating the rotor of a straight bladed vertical axis wind turbine in fixed geometry, using stalling of the blades to control
power, gives a less severe loading than had been anticipated and

demonstrates good power control characteristics. As a result of this and parallel design studies, it has been shown that savings up to 20% in the cost of energy may be obtained from a fixed geometry arrangement. It has been decided that for future machines this arrangement should be adopted. Rather than repair the damage to the rotor of the 17m diameter wind turbine on the Scilly Isles, the rotor is being redesigned to have a fixed geometry configuration. The central part of the rotor crossarm will be retained but the outer parts of the crossarm and the blades will be replaced. It is expected that the wind turbine will be fully reinstated in the summer of 1988.

5. PROGRAMME

To allow time to demonstrate the fixed geometry rotor it is expected that the project will now be completed in March 1989.

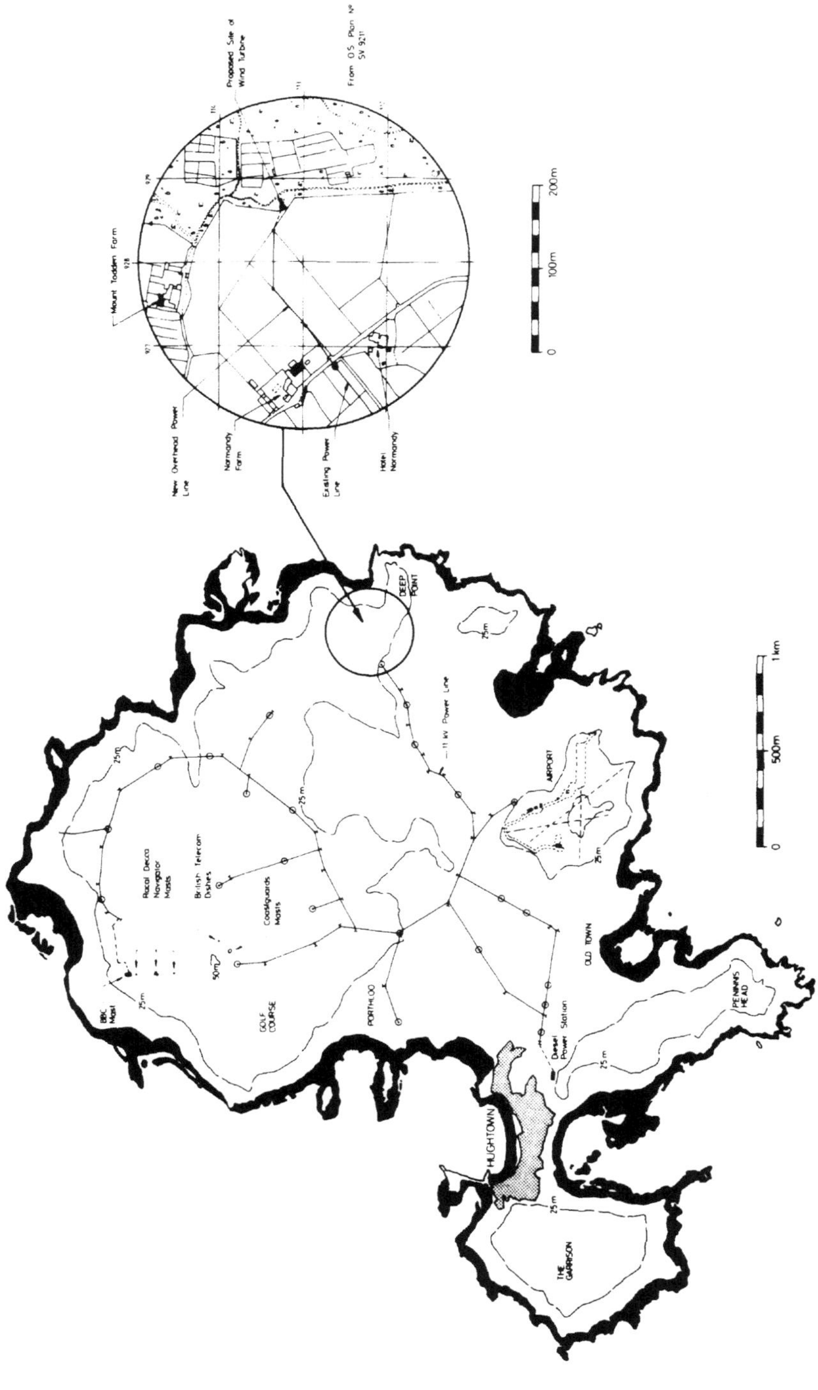

Figure 1 St Mary's, Isles of Scilly

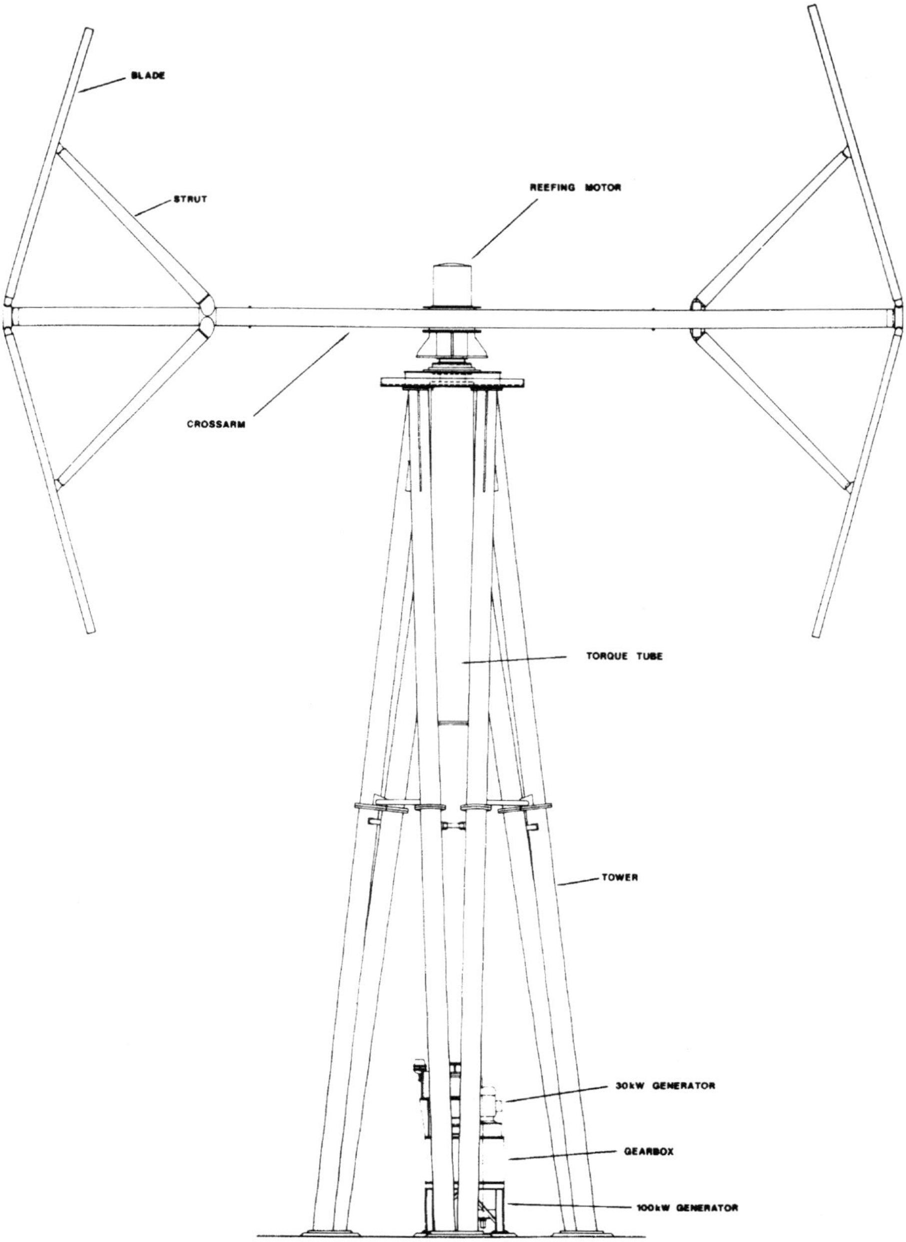

Figure 2 General Arrangement of 17 Metre Diameter VAWT

Figure 3 17 Metre VAWT on St Mary's

Project Nr 403/83 HE

KARPATHOS ISLAND WIND PROJECT

G.VERGOS, J. TSIPOURIDIS, A. ANDROUTSOS, P. PLIGOROPOULOS
A. KORONIDIS
P.P.C. DEME, NAVARINOU 10, ATHENS 106.80 - GREECE

Summary

The installation of a 175 KW pitch controlled W.T. in Karpathos, which feeds the island's small grid (diesel station, 4MW installed capacity) is intended to prove the technical and economic viability of such a scheme, where the degree of penetration of the W.T. will be critical, as the island's demand is highly variable with extreme low values during the winter period. The project was delayed due to siting and access problems and was finally commissioned in February 1987. Its operation so far has not been quite satisfactory as problems associated with the W.T. control/operation and data recording have hampered our efforts to establish a proper operating status. Finally operation ceased altogether recently due to a blade throw.

1. INTRODUCTION

The aim of this project was to establish the reliability of operation of a medium size wind energy conversion unit connected to a small grid powered by the island's diesel station. The fact that the machine was on the upper medium size range (175KW) and the island's diesel station's installed capacity was only 4MW, with very low demand values for the winter period (as the island is a tourist spot with peak electricity demand during the summer period), meant that the parallel operation of the two energy sources would produce very interesting results.

The innovative feature for PPC was the fact that the W.T. was pitch-controlled and its operation was going to give us the required experience from such a unit, in view of our future programmes.

There were certain difficulties in site selection as the originally chosen one had to be abandoned and a new one had to be selected. There were further problems associated with civil works as local contractors were hard to find and the required concrete and steel qualities were not available locally.

2. DESCRIPTION OF WT'S COMPONENTS

The W.T.G. is an horizontal axis, upwind, pitch regulated machine produced by HMZ-Windmaster (Belgium). The main W.T.G data are as follows :

Nominal power	175Kw at 14,5 m/s
Cut in wind speed	5 m/s
Cut out wind speed	25 m/s
Number of blades	3

Rotor diameter	21.8 m
Hub height	23 m
Cone angle	0°
Tilt angle	4°
Tower height	22 m
Tower diameter (base-top)	1,96m/1,08 m
Rotor speed	52 rpm
Generator speed	1523 rpm
Voltage	3 X 380 V
Frequency	50 Hz

In detail the description of WTG's components is as follows :

Rotor : The rotor has 3 tapered cantilevered blades made of Polyester Reinforced Fiberglass. Each one is 10,5 m long with a twist of 10,8°. The airfoil is NACA 643-618

Gearbox A P.I.V. speed-increaser of 290 KW rated power is installed between the main shaft and the generator (ratio 1/27.8).

Generator A BBC 3 phase asynchronous generator is used. Protection class is IP 23 and insulation class is F.

Brake A drum brake is fitted into the main shaft in order to secure the rotor during service and maintenance works in the nacelle (static torque is 40 Kpm).

Yaw system : When the W.T. is connected the turbine is in free yaw. If the windvane indicates a change of direction for more than a specified time (changeable by keyboard) an active yaw will occur and the turbine will go back into free yaw after it has adjusted (by means of a hydraulic yaw drive) the yaw position.

Tower : The tower is 24-edged, tubular tapered steel, in 2 sections. A door is fitted in the bottom section. Access to the nacelle and service platform is possible by an inside ladder.

Control panel : The control panel is placed in the bottom section of the tower. It is linked to the microprocessor and is provided with a display indicating all the operation data as well as malfunctions and failures. There is automatic start-up after a grid failure. Capacitors are installed to keep cosφ 0.92-0.96. There is also over voltage protection against lightning.

Safety control : All the safety policies lead to a shutdown procedure. There are four types of shutdown :

1. If a minor fault or malfunction occurs there is a normal shutdown by pitching back the blades until the generated power is almost zero KW ; after that the computer opens the main contactor.
2. If a major fault occurs there is an emergency shutdown. Then the computer opens immediately the main contactor and starts pitching back the blades with the safety valve (only spring force).
3. If the rotor revolutions are too high after pitching back during a certain period there is a Blade Jam . Then the computer activates the hydraulic valve so that we get not only the

force of the springs but also the force of the hydraulic cylinder.

3. PROJECT MANAGEMENT

Due to the site change and the time required to establish a new one, the project was delayed at the start. However civil works and grid extension works were completed by Autumn 1986. Simultaneously the W.T. tower which had been constructed at the Greek shipyards of Skaramaga had been delivered. Thus when the machine arrived in Jan. 87 and following custom clearance everything was transported on site and erection was accomplished in just over a day. Commissioning took place on the 17.2.87.

The operation and difficulties encountered in the first year are presented in chapter 5. An important activity that was delayed due to its innovative character and the development work required was that of the design, purchase, testing and installation of the Data Acquisition system by PPC personnel.

The manufacturer HMZ, provided a microprocessor based control system, in order to control and regulate the W.T. The system measures KW, wind speed, wind-direction, blade-position, yaw position and r.p.m.

The VI-8088 control system, according to manufacturers documentation can also support through an RS 232 port, asynchronous communication to a Hewlett Packard type 86 microcomputer.

The data communications protocol for message exchanges is of the following type.
CONTROL B, ADDRESS, CONTROL REPLY, ASCII DATA (49) CHECHSUM, CONTROL C.

The following Data (49 ASCII characters in total) are transmitted according to HMZ's standard protocol.

DATE (Year-month-day-hour-minute)	10 bytes
KW Sec (generator or motor)	4 bytes
Windspeed	3 bytes
Winddirection	2 bytes
Blade position	3 bytes
Yaw position	5 bytes
RPM	4 bytes
Current operating mode	4 bytes
Disconnecting reasons	4 bytes
Imput status	4 bytes
Faults	6 bytes
	46 bytes

According to HMZ supplied information the acquisition system and communication protocol can be expanded to include further information according to PPC's requirements.

The configuration (fig.1) utilizes extensively the data acquisition system provided by the wind-turbine manufacturer to avoid equipment duplication and achieve cost reduction. Hewlett-Packard equipment is used for compatibility to the existing equipment and for improved reliability since the wind turbine/data acquisition and storage system have been installed in a remote area and harsh enviromental conditions. An HP-86B computer with a 128 Kbytes of CPU memory is used. This computer was connected to the VI-8088 controller (being supplied by HMZ) via an RS-232 asynchronous data interface. This interface is supported by the HP 0087/15003 I/O ROM in the HP-86B computer.

The Hewlett-Packard HP 9121 D dual 3,5" discette drive, with a storage capacity of 540 Kbytes (270 Kbytes/disc), is used.

Data acquisition is done in a cyclic wav at a scanning interval of the order of seconds (i.e 10 seconds). Active and Reactive power measurements are integrated over this interval to provide energy data KWh, KVARh.

An averaging subroutine is used to calculate the average value for each measurement during 10 minutes. This value is stored for each measurement in the microcomputers RAM.

The basic principles of the data storage scheme which is used are given below.

All data are written in ASCII code. Since the comma (","=ASCII (44)) is used as a field separator, all data are presented in free-field format without the necessity of keeping leading zeros or spaces, as signed integers or decimal fractions, with the stop (","=ASCII (46)) used as the decimal point. If any field in the format of a record is not applicable or data are not available for it, it is indicated as an empty field by its final comma following directly the preceding on. However, all commas directly preceding the end-of-record are suppressed. Thus, if in data record 2 only the 4-th and the 6-th item is available, that record is written as :

2,,,,, item 4 ,,,, item 6´ CR´

A periodic data transfer (i.e every one hour) is performed to transfer data from the microcomputers RAM to the mass storage device, the 3,5" disc drive. Thus, the continous operation and consequtive wear of the disc drive unit is avoided.

The D.A.S. was installed on the 7.2.88 and results will be available for the next meeting.

4. RESULTS OF OPERATION

The overall operation of the W.T. to date has been hampered by a number of problems which make difficult the overall assessment of this project.

To start with the delay in the installation of the data acquisition system (which was finally completed on 7.2.88) resulted in the lack of analytical operation data. This in effect meant that problems occuring could not be properly evaluated. Furthermore the remoteness of the site and the absence of a remote control system (which is expected to be installed by the summer) led to delays in taking corrective action whenever it was required.

More specifically, from commissioning (17.2.87) to mid summer (12.7.87) there was a significant discrepancy between output read out and actual KWh produced (as evaluated at the diesel station) by about 40%. The problem was due to an electronics fault that was partly corrected in July.

Following that and up to January 1988 the W.T. operation (from commissioning) was characterised by too frequent stops which were primarily attributed to high frequency and gust winds and clearly a combination of the two.

The repairs and maintenance which were carried out in January 1988 dealt with all these problems, in an effort to put an end to the situation as it had evolved. Thus the power transducer responsible for the data read out was replaced, the frequency eprom was changed, so that the frequency range was now 48.5-51.5 Hz from 49-51 Hz, and finally the pitch sensor which was discovered faulty was also changed. However very shortly after the repairs the pitch sensor failed again and the W.T. was stopped.

Finnaly while we were waiting for the replacement, the W.T suffered a vital blow. On the night of the 10.3.88 one blade was destroyed. The blade was torn off near the hub, was sliced along its long axis and the pieces were found at a distance of about 15m to 40m respectively.

Subsequently the W.T. was inspected by PPC personnel (13.3.88) and an HMZ engineer (31.3.88). The inspection further revealed that the yaw pinion was also damaged.

Therefore we are now expecting the replacement of all three blades by new ones, as well as the replacement of the yaw pinion.

In conclusion it can be said that a series of events which happened one after the other by coincidence alone, led to a disrupted operation. It is hoped that following the repair works due in May 1988 the machine will operate normally achieving thus its original objective. In spite of all the problems total production for a year (Feb.87 - Feb 88), was 203.000 KWh.

Operation results and other data are shown in figures 2 to 8.

5. DEGREE OF SUCCESS AND OUTLOOK

Obviously the problems encountered in the first 13 months of operation do not allow for an optimistic appraisal of the project. However one cannot avoid the thought that on top of all the problems associated with the site, access, the small grid, the gusts etc, the project ran into a number of coincidences that never let it get off the ground.

One can only expect that following replacement and repair work in the next months, operation will be a lot smoother, taking into account the installation of the data acquisition system and the expected remote line connection in the summer.

K A R P A T H O S   I S L A N D

F I G U R E   1

A. SYSTEM FOR DATA ELABORATION AND ANALYSIS

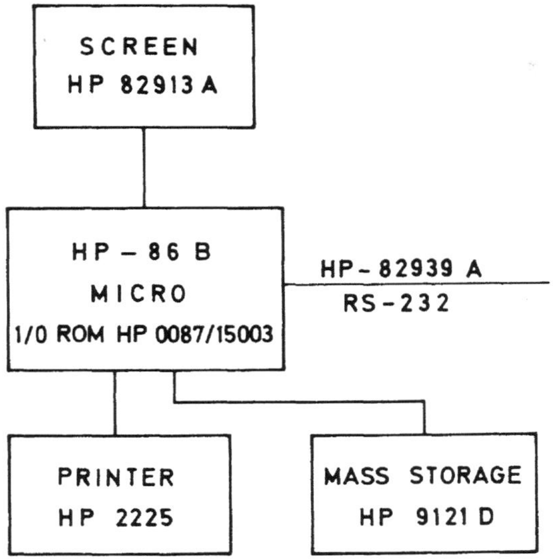

B. DATA ACQUISITION BLOCK DIAGRAM

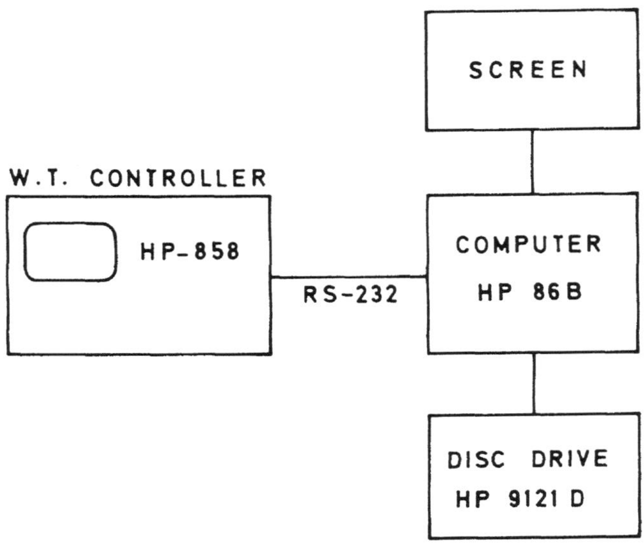

KARPATHOS ISLAND

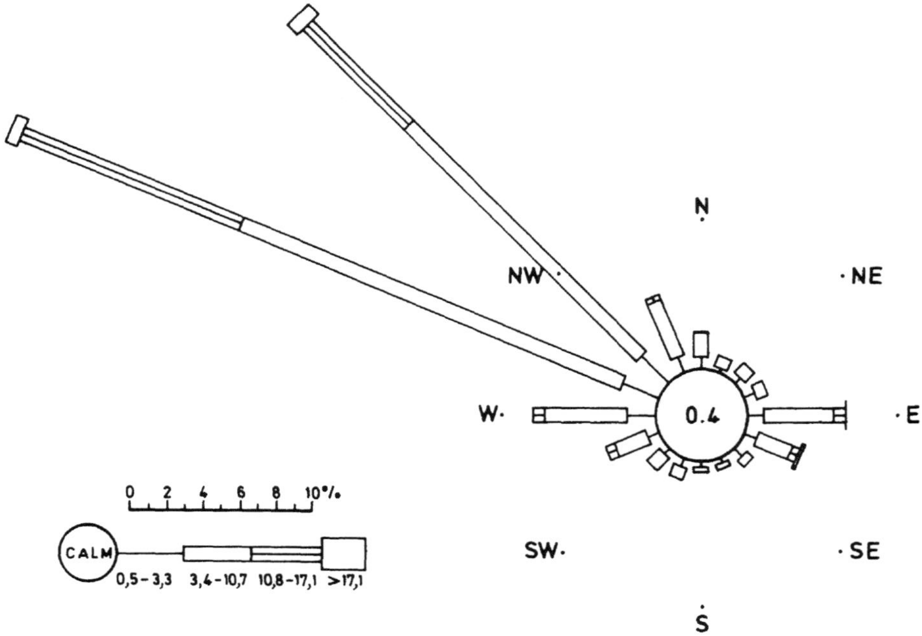

Figure 2. WIND ROSE (May '87 - Feb. '88)

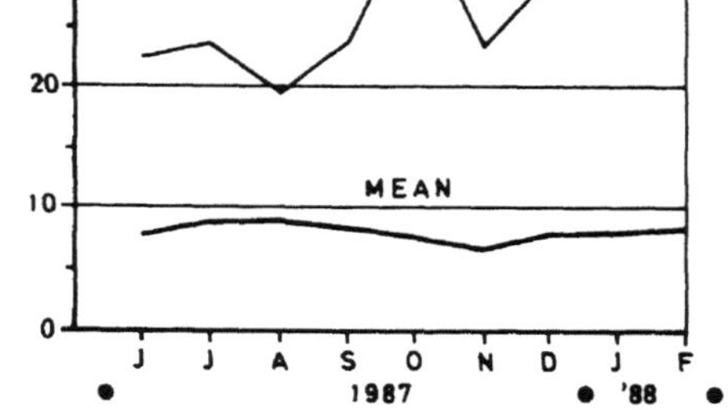

Figure 3. MEAN WINDSPEED - MAX. GUST (1 min average)

KARPATHOS ISLAND
DIESEL POWER STATION

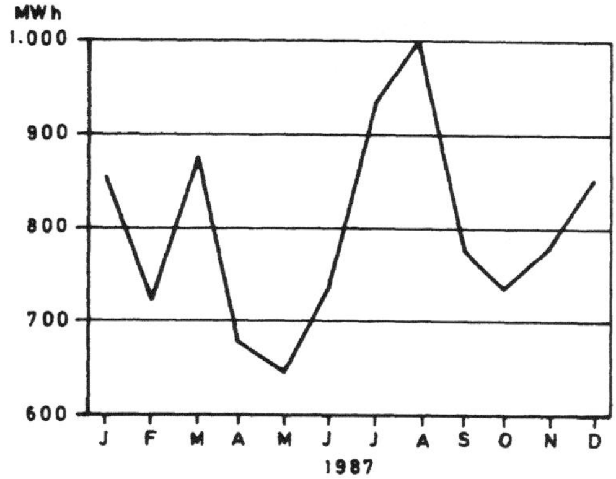

Fig. 4. MONTHLY TOTAL PRODUCTION

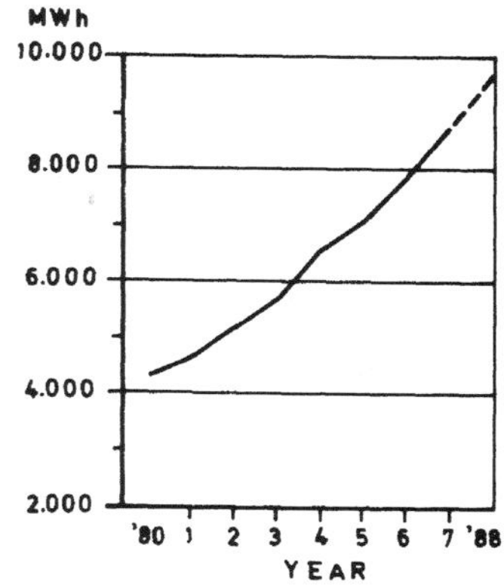

Fig. 5. ANNUAL TOTAL PRODUCTION

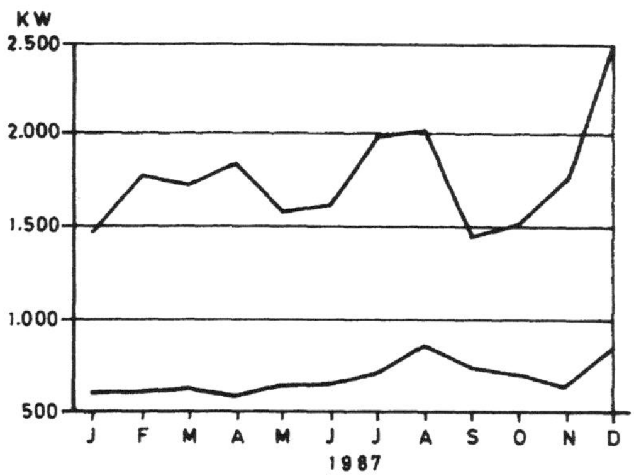

Fig. 6. MONTHLY MAX. AND MIN. LOAD

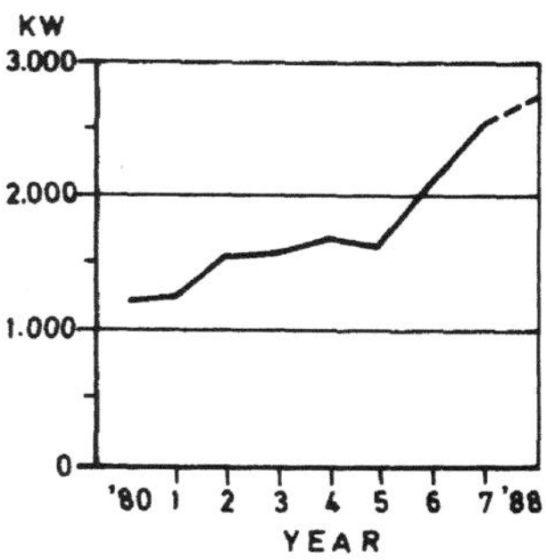

Fig. 7. ANNUAL MAX. LOAD

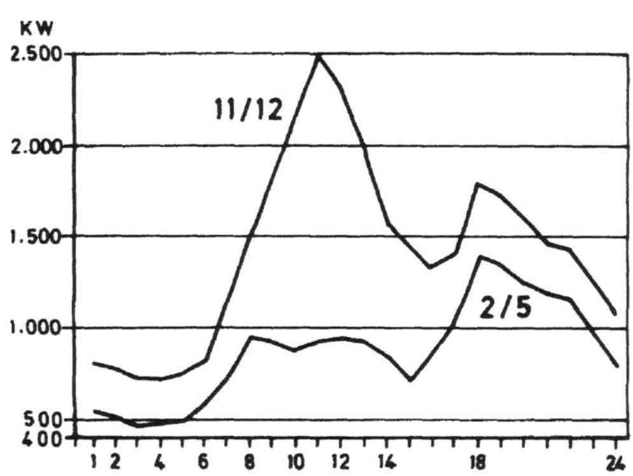

Fig. 8. DAILY LOAD CURVE

Project Nr 476/84 UK

THE SHETLAND WIND DEMONSTRATION PROJECT

G.A. ANDERSON - North of Scotland Hydro-Electric Board
and
D. PASSEY - James Howden & Co Ltd

SUMMARY

The North of Scotland Hydro-Electric Board wind demonstration project with a 750 kW, 45 m diameter James Howden & Co Ltd Aerogenerator has progressed to the stage where construction is nearly complete. The blades are scheduled for installation to meet commissioning the unit in May, 1988.

Environmental considerations have been an important element of the project with local people requiring assurances and undertakings from the Board on visual impact, noise, electromagnetic interference and construction works.

The techno-economic success of the project will be evaluated during the operational phase which includes a two year period for monitoring the performance and structural behaviour of the unit.

1. INTRODUCTION

The North of Scotland Hydro-Electric Board are responsible for the generation, distribution and sale of electricity in the North of Scotland and island groups including Orkney, Shetland and the Western Isles which covers about 25% of the land mass of Britain and 2% of the population. The mainland of Scotland is served by the National Grid with electricity generated by conventional plant. In 1973, when OPEC virtually quadrupled the price of oil, the island communities of Shetland, Orkney, Western Isles and Tiree were supplied by electricity generated by local diesel stations. The oil price rise resulted in the cost of generation on the islands becoming two to three times more expensive than on the mainland. The Board therefore developed a strategy to seek out alternative means of reducing dependence on oil on these island groups (1). Since 1973 Orkney and Tiree have been connected to the National Grid by submarine cable and work is proceeding to connect the Western Isles. In the case of Shetland however greater distances are involved and connecting the island by submarine cable is a longer term prospect.

The Board examined alternative sources of energy including peat, wave energy and wind energy. With the high and persistent wind levels experienced in the West of Scotland and particularly on the islands wind energy emerged as the most promising option for Shetland. The Board, encouraged also by the experience gained with medium sized wind turbines in Orkney, developed proposals with James Howden & Co Ltd for the installation of a 750 kW aerogenerator in Shetland. The objective of the proposal being to determine whether wind energy would provide a cost effective means of reducing dependence on oil consumed at Lerwick diesel power station and to evaluate the potential for wind farms. The proposal was submitted to the Commission of European Communities and was awarded their financial support under the programme for wind demonstration projects.

The project involves the construction of the aerogenerator, foundation, access road, connection to the 11 kV distribution system, provision of remote control equipment and monitoring the operation of the wind turbine generator for a period of two years.

The development provides for an aerogenerator to be demonstrated with a 45 m diameter rotor on a high energy hilltop site with mean wind speeds of over 10 m/s and a synchronous generator of the type which could be used for wind farm developments on Shetland.

The aerogenerator is being constructed at Susetter Hill on the mainland of Shetland and will be controlled remotely from the Board's diesel power station at Lerwick which is 30 km from the site.

2. WIND TURBINE GENERATOR

The HWP 750/45 Wind Turbine is a three bladed, horizontal axis, fixed pitch, upwind machine, which will produce a net 750 kW at a rated windspeed of 13 m/s. The power - wind speed characteristic curve is shown in Figure 1.

The rotor diameter is 45 metres and the hub height of the machine is 35 metres. Figure 2 shows the general arrangement of the machine.

The tower is free standing, fabricated in steel and largely cylindrical in section. Towards the base, the tower flares in order to increase the foundation bolt pitch circle diameter, thus spreading the loads transmitted from the machine to the foundation and to provide accommodation for the necessary electrical and control equipment.

The three bladed upwind rotor is employed for its properties of smooth power output, good start-up characteristics and reduced dynamic blade loads.

Rotor blades are constructed from a wood/epoxy composite which exhibits a high strength to weight ratio and excellent fatigue properties. Fixing to the spheroidal graphite cast iron hub comprises high tensile studs set in a carbon fibre/epoxy grout. Stop and start control of the rotor is achieved by means of moveable blade tips, which also provide the means of power control at wind speeds greater than those required to produce rated power output. A hydraulic actuator mounted on the end of the main blade rotates the tip section into the run position. The blade tip section swivels about a compound spar mounted radially in the end of the blade and is biased towards the 'Stop' position by a pre-loaded spring. This arrangement provides a fail safe facility whereby loss of hydraulic pressure in the actuator permits the tip to revert to the 'Stop' position. The blades are designed and manufactured entirely by Howden in their new blade manufacturing facility opened in 1987.

The arrangement of machinery and equipment in the nacelle is shown in Figure 3.

The rotor hub is bolted to the low speed forged steel shaft, which in turn is mounted in two, grease lubricated, self-aligning, split roller type bearings. The shaft also has a central axial hole to provide a path for the supply of hydraulic fluid to the blade tip actuators.

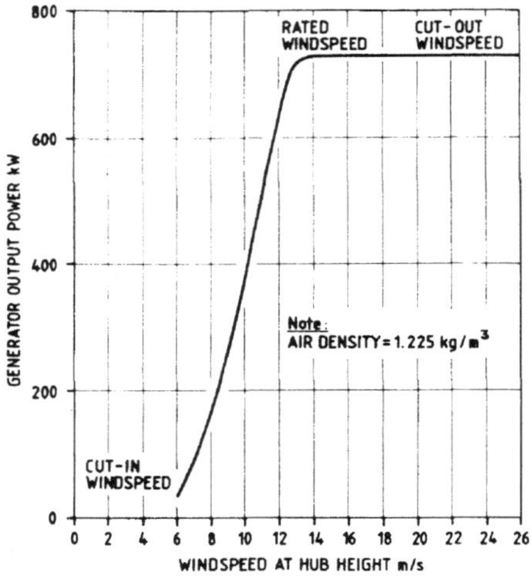

Figure 1 : 750 kW AEROGENERATOR POWER WINDSPEED CURVE

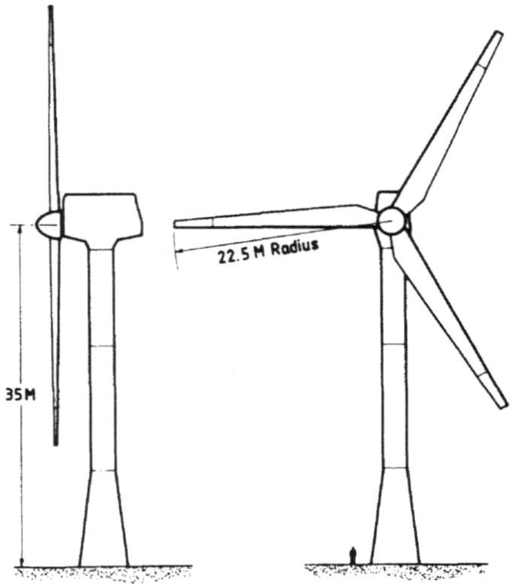

Figure 2 : GENERAL ARRANGEMENT 750 kW AEROGENERATOR

Figure 3

The two-stage, speed increasing gearbox comprises a first stage planetary and second stage parallel shaft arrangement, which increases the blade rotational speed of 30 rpm to a generator rotational speed of 1000 rpm via a high speed transmission shaft and fluid coupling. The latter feature introduces compliance into the system to attenuate gust effects on power output.

The generator itself is an 11,000 volt, 50 Hz, 1000 rpm synchronous unit with a drip-proof enclosure and a marine finish to withstand the salt laden environment on the island site. The automatic voltage regulator and power factor controller are mounted in the unit switchgear which is housed in the base cone of the wind turbine tower. In the case of this installation, the auxiliary supplies for the machine are being taken from a pole mounted 11 kV/415 V transformer provided by the North of Scotland Hydro-Electric Board. However the facility to include an auxiliaries power supply transformer within the switchgear is available.

The wind turbine control system consists of two standard Programmable Logic Controllers (PLC), one located in the nacelle and the other at the base of the tower.

The base PLC as master controller is responsible for monitoring all machine conditions electrical and mechanical, in order to allow start-up or to initiate a shutdown. It is also responsible for adjusting the generator voltage during run-up, and for operating the main power contactor when the generator speed, voltage and frequency are within limits.

The nacelle controller when instructed by the master controller looks after the operation of the shaft brakes, the yawing and hydraulic functions. It also controls the operation of the rotor from acceleration up to synchronous speed, power regulation when generating, and run down when a shutdown is initiated.

Remote control facilities are provided to allow the machine to be shutdown, the power setpoint to be adjusted and for the generator to be switched from power factor control to voltage control to allow remote adjustment of the generator voltage. Remote monitoring of the machine is also available for the remote station to monitor the condition of the machine. The remote control facility is performed via a microwave link from the North of Scotland Hydro-Electric Board generating station at Lerwick.

Rotor braking is provided by means of a steel disc mounted on the low speed shaft with four spring applied, hydraulically released calipers, which use a reduced braking force for normal service stops and full brake capacity to halt the rotor in the minimum time for an emergency stop. A mechanical locking pin is inserted through the brake disc and mating bracket to provide positive immobilisation of the rotor and drive train during maintenance operations.

The nacelle bedplate is connected to the top tower flange via a standard slewing ring, with angular positioning of the nacelle assembly relative to the wind direction being achieved using a combination of geared hydraulic motors and caliper type brakes which run on the machined tower top flange. Again, the brakes are fail-safe, spring applied, hydraulically released and a locking pin is provided for maintenance duties.

The drive train and hydraulic power pack are all mounted on the nacelle bedplate, which is fabricated from structural steel throughout.

The nacelle cover which provides the machinery on the bedplate with protection from the elements is manufactured in GRP, with apertures providing the necessary ventilation for equipment and for hub access and lowering equipment to ground level, by means of the electrically powered integral hoist. Internal lighting is augmented by the inclusion of translucent panels in the roof.

Considerable emphasis was placed upon achieving a pleasing visual presentation of the machine when arriving at the design of the nacelle cover shape. This was both to maintain the Howden reputation for aesthetic appeal set by the earlier 300 kW machine on Orkney and in recognition of the environmentally sensitive nature of the selected site on Shetland.

To complete the consideration for its place in the landscape, the external colour of the tower, nacelle and blades is a very pale grey as recommended in the Environmental Assessment Report prepared on behalf of the North of Scotland Hydro-Electric Board.

Consideration of the remoteness of the site and means available for transport and erection were taken fully into consideration during the design of the unit. This has enabled the use of conventional road vehicles for the journey to site and embarkation and disembarkation from the commercial roll-on/roll-off ferries which operate between the mainland and Lerwick.

Similarly, the road access constructed from the public highway to the turbine foundation by the North of Scotland Hydro-Electric Board was kept to economic widths and gradients.

The site erection of the tower sections and nacelle assembly was achieved with a single 300 tonne capacity crane, the largest single lift comprising the nacelle assembly being approximately 55 tonnes.

The tower stands on a reinforced concrete gravity type foundation constructed as part of Howden scope of supply. It was not possible to take advantage of keying the block into bedrock, due to the weathered nature of the rock encountered during the site soil survey. Spoil removed during construction was partly transported from the site and partly deposited in deep peat hags for subsequent landscaping and re-seeding.

It is planned to complete commissioning of the wind turbine generator during the Spring of 1988, at which time a two year monitoring programme for the machine will commence which will include operation, performance and structural monitoring under the widest possible combinations of operating conditions.

The annual energy capture of the machine is approximately 3270 MW hrs when based on an annual mean wind speed of 10.4 m/s and an availability of 100%.

3. PROJECT REALIZATION

The Board's Project Management structure is provided by the Technical Director and Project Co-Ordinator supported by specialists. The Board define and implement the contract strategy with co-ordination of contract interfaces, programme and budgetary control. The Board commissioned the services of environmental consultants and quality assurance specialists to assist meet project objectives.

The Board have been responsible for the construction of the access road, electrical distribution and telecontrol/communications systems. Sub contractors are chosen from competitive tender except where particular requirements dictate eg compatibility with existing works. Where expertise exists on Shetland then local organisations are invited to quote. The access road and some welding operations are examples of work undertaken by local firms.

The main problem encountered in the project related to securing a site for construction. The Board's first choice was to install the machine at a hill named Scroo Hill. After long protracted negotiations this proposal was abandoned due to conditions imposed by local peat cutting interests involving high access road costs. In order to progress the Board decided to apply for consent to install the machine at their second choice site at Susetter Hill. This application was made in November 1985. The Board were required to provide an Environmental Assessment Report on the Development. Following consultations and a local public meeting consent was finally received in June 1986. Subsequent negotiations for land acquisition were concluded in August 1986. The period taken from initial approaches to develop aerogenerators in Shetland to final approval spanned a period of nearly 2 years.

The Environmental Assessment Report (2, 3) addressed facets of the development identified by the Director of Planning, Shetland Island Council and a petition was raised by local people objecting to the proposal. The main areas of concern emerged as the visual impact of the aerogenerator and its prominence on a hilltop site, noise and electromagnetic interference. As described earlier, James Howden recognised the need for the aerogenerator to be visually pleasing and engaged consultants to assist with the aesthetic design. With regard to electromagnetic interference, the British Broadcasting Coporation (BBC) were commissioned to advise on the effect of the aerogenerator on the District. The local community were outwith recognised television reception areas and had installed a self help device to facilitate TV reception. Field tests were carried out by the BBC and their report identified households where TV reception would be adversely affected (4). To resolve the matter the BBC, with assistance from the Board, accelerated their longer term plans and have installed a relay station in the area. This is supplemented by receivers, transposer and transmitters installed on site by the Board. In the case of noise, information was provided from tests with aerogenerators at Burgar Hill and from the Institute of Noise and Vibration at Southampton. Noise measurements will be taken when the unit enters service.

Problems during aerogenerator manufacture have been minimal. Dimensional errors on the gearbox annulus casing resulted in a replacement forging being required. Restrictions in transportation resulted in one of the tower sections requiring a joint to be welded at site prior to erection. A problem arose in transporting the 300 tonne capacity erection crane to site when a minor road started to subside and break up under the load. A one week delay was encountered whilst Howden arranged for road modification after which the crane vehicle was successfully driven onto site.

4. COMMISSIONING AND OPERATION

Work is proceeding towards commissioning the aerogenerator in May, 1988. A meteorological study conducted under a collaborative project with the Board, Rutherford Appleton Laboratory and Strathclyde University has provided quantitative information on the wind characteristics. The study includes measurements from a 45 m meteorological mast and instrumentation installed at site. Information on turbulence intensity, wind shear, Weibull coefficients and wind speed have been used in the design of the machine (5, 6, 7). The annual energy capture of the machine is estimated as 2940 MWh when using an availability of 90% for the aerogenerator and the 2 year contract monitoring period will be used to validate the design and performance predictions.

5. PROGRAMME

The programme is for the machine to be commissioned May, 1988 and for this to be followed by a two year monitoring period.

The programme delays were principally those relating to securing a site for construction as discussed under item 3. In addition there has been some delay as a consequence of Howden setting up a new factory in Southampton for in-house blade manufacture and an initial high work load in this factory. Notwithstanding this, blades are at the final stages of manufacture for delivery to site by early May 1988.

6. COSTS

The investment cost for the project will be determined when all costs for the work have been received. The original estimate at £1,417,000 has not been revised. This cost includes the aerogenerator, foundation, access road, electrical distribution system, remote telecontrol and telecommunication systems. In addition the cost includes for a 2 year monitoring period at £125,000. The CEC contribution to the contract is £470,800.

Evaluations based on a 5% discount rate, 1% maintenance costs and 20 years life give a cost of electricity produced at 3.92p/kWh if monitoring costs are omitted, ie a capital cost of £1,292,000.

On the same basis, but using the capital cost to the Board as being £1,417,000 less the CEC grant, then the cost is 2.98p/kWh.

The above compares with the cost of fuel saved over the life of the project currently predicted at about 3p/kWh. The current cost of fuel is, however considerably less than this.

It should be noted however that these are nominal values and that full investment appraisal by the Board provides for sensitivity analysis of economic factors affecting cost and for scenarios with a submarine cable connecting Shetland to the mainland grid within 20 years.

The potential reduction in cost for series-produced aerogenerators would be influenced by manufacturers commercial consideration. ETSU Report 30(8) suggests that for each doubling of production then cost reduces to say 85-90% of the previous value. On this basis costs in the range £700 to £900/kW installed are derived for series-produced systems, when making an allowance for site costs.

7. EVALUATION OF THE PROJECT

The contract works will be evaluated against project objectives from information determined during the monitoring period and when final capital costs are established.

Techno-economic factors monitored for evaluation of the project costs will include appraisal of technical life from details of structural behaviour and fatigue; availability; operating and maintenance costs; performance and annual energy capture.

The economic benefit of the project relates to the cost of fuel oil saved. In this regard volatility in the price of oil will present difficulties in predicting the benefit. Since contracts were placed the price of crude oil has fallen from about $30 to about $15 a barrel and in 1986 could be purchased for less than $10 a barrel. As the economic benefit from wind energy is directly proportional to the cost of oil this fall in price has placed further pressures on the wind turbine industry to reduce costs.

8. MARKET EXPLOITATION

The North of Scotland Hydro-Electric Board have supported the development of wind energy. The 300 kW prototype aerogenerator installed at Burgar Hill, Orkney by James Howden & Co Ltd has led to nearly 100 similar machines being installed, mainly in overseas markets. It is expected that the 750 kW machine will meet similar market interest as the process of increasing the size of machines continues to meet utility requirements.

REFERENCES

1 VERNON, K R, Future Prospects for Hydro Elecricity and Wind Power. Proceedings of the Royal Society of Edinburgh, 92B, 107-117, 1987.

2 CAIRNS, W & J (Environmental Consultants), Report for the North of Scotland Hydro-Electric Board, Susetter Hill Aerogenerator, Shetland. Produced by W & J Cairns.

3 MILLER, J S, Environmental Assessment of Wind Turbine Generator Project, Susetter Hill, Shetland. Paper presented at the Ninth BWEA Wind Energy Conference, 1987. (Proceedings yet to be published).

4 BATE, P, An Assessment of Interference to UHF Television Reception from the proposed Wind Turbine Generator on Susetter Hill, Shetland and Relay Station Considerations, BBC Television Service Planning Note, 1987.

5 BOSSANYI, E A, HALLIDAY, J, GARDNER, P, Analysis of Wind and Turbulence Measurements on Shetland. Proceedings of the Eighth BWEA Wind Energy Conference, 1986. Edited by M B Anderson and S J R Powles. Published by Mechanical Engineering Publications Ltd.

6 GARDNER, P, ANDERSON, G A, et al, Wind Integration Study for a Mediumscale Autonomous Electricity System. Proceedings of the EWEA Conference, 1986. Edited by W Palz and E Sesto, ISES, Rome.

7 TWIDELL, J A, ANDERSON, G A, et al, Wind Generated Power for Shetland: Tactical Planning for the 30 MW Peak Autonomous Grid and Diesel/Thermal Plant. Paper presented at the Ninth BWEA Wind Energy Conference, 1987. (Proceedings yet to be published).

8 Energy Technology Support Unit for the Department of Energy, ETSU Report 30, Prospects for the Exploitation of Renewable Technologies in the United Kingdon, Her Majesty's Stationery Office, 1985.

Project Nr 626/84 HE
A 100 KW WIND TURBINE SYSTEM CONCEPT

G. Bergeles and N. Athanassiadis
Nat.Techn. University of Athens

Summary

This report describes the concept and final design phases of the EEC DGXVII contract. Under this contract and with the consponsorship of EEC/DGXVII and the Greek Public Power Corporation (PPC), the laboratory of Aerodynamics of the Nat. Techn. University of Athens and PPC undertook the design and construction of a 100 KW horizontal axis wind generator (HAWG); the machine is designed to operate on the island of Skyros at a site where wind measurements indicate a mean annual wind speed of 6.5 m/s.
The features and characterestics of the wind generator as they evolved from a series of trade off studies include an upwind variable speed rotor of 20 m diameter with three fixed pitch blades, a planetary step-up gear box, a hydraulic low speed shaft disc brake, an electric high speed shaft disc brake and an 175 KW electric motor of 1500 RPM. Sismic analysis of the HAWG and its foundation indicated that an earthquaqe loading is of critical importance to the structural integrity of the system.
The final design package of the machine consists of the components specifications, the anticorrosion measures due to sea proximity and of approximately 100 construction drawings; the Greek Aerospace Industry or BIEX probably will be the main subcontractor for the construction of the machine.

1. INTRODUCTION

Ancient Greeks had a King, named AIOLOS, who controlled the Winds in these areas and helped sailing among the 2000 small and big islands in the Meditterenean sea. AIOLOS is still alive (but as a Wind) and has made the Greek islands rich in Wind Energy Potential which modern Greeks and their EEC partners are challenged to harness. It is estimated that with approximately 2000 Wind Generators of 50 meters in diameter installed all over Greece, a good 25% of the electricity consumption of Greece can be covered by the production of these Wind Machines at competitive prices; looking particularly to the high wind energy potential of the Greek islands and supposing that the Cyclades islands could be connected electrically to the mainland these islands could form the Wind Power stations for whole Greece as the thermal power stations are today in the mainland. In order to materialise the harness of the Wind Energy Potential the development of technology of Wind Generators is essential and the current EEC contract aims as this goal; a 100 KW HAWG is considered an optimum size machine with economic value for the isolated electric grid of the Greek islands. Therefore this contract undertook the development, the construction and monitoring of 20 meters diameter three bladed upwind HAWG.
In the next paragraphs details of the design choices will be presented with particular emphasis on the novel features of the Wind Generator under construction.

2.SYSTEM DESCRIPTION

This section describes the OA/100 KW Wind Turbine Generator at the final design stage before construction. The final design evolved after preliminary trade off studies as regards subsystem reliability and current experience from the operation of machines of similar size. The main target was to develop a machine of high structural reliability and performance as in detail is presented in the following paragraphs.

2.1. General arrangement and characterestics

The general arrangement and characterestics of the OA/100 KW Wind Generator at its current configuration is shown in figure 1. The Wind Generator is designed for operation at sites where the mean annual wind speed is around 6.5 to 7 m/s at 10 meters height. The system starts generating electricity at about 4 m/s and delivers its rated 120 KW electric power at wind speeds of 12 m/s (at hub height). The machine is shut down at 25 m/s to avoid operation at high loads. The expected annual energy output is around 200000 KWHs. During operation the wind turbine is connected to the power grid through transformers; capacitors and filters are used for assuring a high reactive power and an acceptable sinusoidal voltage form.
The Wind Generator is a horizontal axis machine equipped with a 20 meter diameter upwind rotor. It has three blades rigidly attached to the rotor. The rotors centre of rotation is 24 meters above ground level. A 175 KW electric generator is driven via a step-up planetary gear box with a flexible coupling connection. The generator, the gear box, the hydraulic system, the disc brakes and other support equipment are housed in a nacelle mounted at the top of a cylindrical steel tower. The nacelle can be yawed to keep the rotor oriented correctly into the wind as the wind direction changes. The rotor and consequently the generator operate at variable speed controlled by a static inverter; the latter, activated from a programmable logic controller, gives variable frequency in the circuit of the generator but dilivers constant frequency 50 Hz to the grid; The microprocessor is designed to allow unattented operation at a remote site. The microprocessor monitors wind conditions and the operational status of the wind turbine. Equipment failures result in automatic shutdown of the machine.

Rated power (electric)	120 KW
Rotor diameter	20 m
Rotor orienration	upwind
Rotor airfoil	NACA 44XX
Rotor type	fixed pitch
Rated wind at hub	12 m/s
Cut-off wind speed at hub	25 m/s
Rotor RPM	variable from 26 up to 58 RPM
Generator RPM	variable from 670 up to 1508 RPM
Generator type	asychronous - squirt cage
Gear box	two-stage planetary,1:26 ratio
Hub height	24 m
Tower	steel round
Yaw control	electric
Electronic control	programmable logic controller acting on the inverter

2.2. Subsystem design

This section describes the basic subsystems of the Wind Turbine. It is divided into rotor, drive train, nacelle, tower/foundation, electronic control and electrical power system sections. The operation of each subsystem, its function, and the selection of the particular characteristics are presented.

2.2.1. Rotor

The rotor has three blades rigidly attached to the hub; the blades are manufactured by STORK of Netherlands and have been chosen because of their aerodynamic characterestics and their smaller weight compared to other blades commercially available; at the initial design of the project, blades with tip control were envisaged to be used for power control and load reduction. As at this stage no reliable commercially available blades can be found, constant pitch blades were selected; on the other hand since constant pitch blades lead to excess loads on the hub and unctrollable power and loads at high wind speeds it was decided to allow operation of the rotor at variable speed using a variable speed electric generator, so controlling the rotor speed, control of the loads and power could be achieved even though fixed pitch blades are used and in the same time maximize the energy output. The length of the blade is 9.6 m it has a 14 degrees twist from hub to tip and utilises a NACA 44XX series airfoil. The weigth of the blade is 450 Kg and its made of glass-fiber reinforced polyester; if technology develops the blades is envisaged to be equipped with movable tips, centrifugally activated as an extra safety measure for overspeed control.

2.2.2. The Drive Train

The drive train subassembly consists of a low speed shaft with nominal speed of 58 RPM, a low speed hydraulic disc brake, a low speed flexible coupling, the step-up gear box, the high speed electromagnetic disc brake, the high speed flexible coupling and the generator. These major components are shown in figure 2.

The blades are of flanged type to be assembled to the hub. The hub is made of cast steel and it is assembled together with the hydraulic disc brake of 910 mm diameter to the flange of the low speed shaft; the shaft has a total length of 1560 mm with a front diameter of 240 mm and a rear diameter of 160 mm. At this rear part the shaft is connected via a flexible coupling to the gear box.

The gear box is a two stage planetary type with a step-up ratio of 1:26 with in line input-output shafts. The generator is an asychronous generator squiret cage type with nominal electric output of 175 KW. This generator was selected to secure a safe operation since the rotor has fixed pitch blades, for power and loads control and maximum energy capture. Two independently controlled braking mechanisms, one hydraulic and the other electric of the fail safe type have been chosen for safety reasons. The disc brakes are equipped with sensors for early warning of pads wear and maintainance purposes.

2.2.3. The Nacelle

The nacelle houses the major subsystems of the Wind Generator, such as the drive train, generator with its accessories, yaw bearing and drive etc. Its primary function is to provide a rigid platform for the system components, react to rotor loads and provide environmental protection for the components, fig. 3.
The nacelle is supported on welded frame of 3500 x 860 mm dimensions which is formed by two NP136 steel beams which are interconnected by steel plates where the bearings, the gear box and the electric generator are supported. The nacelle sits on a steel tower via special ball bearing (similar to ones used in cranes) and has internal toothing of 840 mm in diameter with 84 teeth. The yawing of the nacelle is achieved via an electric motor via a reduction gear box. A special slipping disc which operates also as brake assures the allignment of the rotor to the wind direction, fig. 4. The control system utilizes a wind sensor to determine wind direction. To allow for the short period wide directional variations common at low wind speeds, the yaw control system uses half minute average to determine wind direction. An additional brake holds the nacelle from inadvertment yawing due to wind loads during no yaw operation and it applies damping during yaw motion.

2.2.4. Tower/Foundation

The tower is made of steel and consists of three cylindrical parts. The base, with a diameter of 2000 mm and a height of 5500 mm, the intermediate part with 1200 mm diameter and height 5500 mm, and the top part of 1000 mm diameter and a height 10 m on which sits the nacelle. The transition from one cylindrical part to the other is made via conical parts of 1500 mm height. The main shaft of the wind generator is at 24 m height. The tower is formed from curved welded plates of 12 mm thickness and sits on reinforced concrete base of deptph 1.5 m and surface 7x7 m^2 through 36 steel rods of 50 mm diameter. The steel rods reach the whole depth into the foundation to an embeded steel flange. The tower has two doors at the SW side (downwind part of the tower for NE prevailing winds). Internally the tower has a staircase up to the top door. Strict specifications have been imposed for the steel construction of the tower and the nacelle base frame as regards welding, rust prevention and construction tolerances.

2.2.5. Electrical Power System

The electric generator is asychronous, squirel cage, three phase 220/380 volts; its nominal power is 175 KW at sychronous speed of 1500 RPM for 50 Hz operation. It is connected to the electric grid of 15 KV via a static inverter and a three phase transformer 380V/15KV of 200 KVA; capacitors for correcting the cosφ and suitable filters for achieving sinusodial form of the produced voltage of 50 Hz are installed.

2.2.6. Electronic control system

The electronic control system provides the sensing, computation, and commands necessary for unattented operation of the wind generator. The electronic control system of the wind generator consists of a programmable logic controler, SIMATIC 115U, of Siemens where all the inputs either from the manual controls or from the sensors are further processed to energise the appropriate action. The monitoring system of the performance of the wind generator as also the wind data are collected and processed by a personal IBM computer with 50 MB hard disc and 1 MB RAM with serial port. The system in case of mulfunction does not affect the operation of the wind generator whose control is only done by the industrial type programmable logic controler.

2.3. System performance and control

The wind generator is designed to operate at variable speed depending on the wind velocity. This is achieved by introducing a static inverter in the electric circuit of the stator of the generator and the transformer; the static inverter changes the frequency of the stator of the generator depending on the wind speed. The programmable logic controler (PLC)is programmed on an algorithm based on the performance of the rotor. With this system it has been preselected that the rotor operate from 26 RPM to 58 RPM. Figures 5 and 6 show the operation diagrams of the wind generator.

a. The W/G is stopped at wind speeds below 4 m/s. At this phase the electromagnetic brake is in operation. The hydraulic brake is open.

b. When the wind speed is for some minutes consistently above 4 m/s the yaw system is activated and then the nacelle is parked. Afterwards the electromagnetic brake is freed and the generator working as motor of

17 Hz accelerates continuously the rotor to 26 RPM a point where for 4 m/s wind speed the aerodynamic torque is maximum. With the increase in wind speed velocity the program controler increases the frequency of the circuit so as always and up to 9 m/s the system operates at maximum power.

c. For wind speeds between 9 m/s and 11.5 m/s the controler keeps the rotor speed to 58 RPM; at 11.5 m/s maximum rotor power of 136 KW is dilivered. At this point maximum power is achieved but not maximum allowable torque.

d. From 11.5 m/s to 12 m/s the programmer reduces the rotor speed to 52 RPM keeping the max power at 136 KW and permitting the increase of the torque of the generator to a maximum allowable value of 25 KNM.

e. From wind speed above 12 m/s the programmer keeps the above maximum torque constant reducing the rotor speed as wind speed increases.The power output is accordingly reduced.

f. At wind speeds above 25 m/s the program controler reduces further the rotor speed and in the same time activates the hydraulic brake; after stop the electromagnetic brake is activated.

2.4. Weight

The weight are estimated based on preliminary design drawings or manufacturer catalogs of the subsystems; so the

a. nacelle (complete with hub, shaft, gear box, brakes, generator etc) weights 60 KN approximately and

b. tower (complete structure) weights approx 100 KN.

2.5. Design environment

To insure the structural integrity of the W/G and its ability to operate in adverse environmental conditions the design environment was defined as shown:

2.5.1. Extreme Wind Conditions

a. Normal operation of W/G below or at rated speed with gust of 35 m/s.

b. Normal operation at 25 m/s with gust of 60 m/s.
c. Operation at 25 m/s with 40% rotor overspeed.
d. W/G stopped and wind speeds of 60 m/s.

Figures 7 to 9 shows the stresses and moments along the blade from the root to the tip; it is found that the out of plane stresses are by far the most critical factor in the design and stresses arising at nominal operation of 24 m/s with a gust factor of 2.5 are almost double those arising with a gust factor of 1.5. Rotor overspeed due to power loss does not lead to excess stresses.

2.5.2. Natural hazzards

Figure 10 shows the response spectrum of the structures for earthquaqes in Greece. Using this spectrum a sismic analysis of the structure was conducted which led to strenthening of the tower. Dynamic analysis of the structures gave that the first three eigenfrequencies are 1.6 cycles/sec,11.4 cycles/sec and 30.6 cycles/sec. It is noticed that the first frequency 1.6 cycles/sec is much higher than the rotational frequency of the rotor of 1 cycle/sec. The natural eigenfrequencies of the structure were also calculated; the first one was found to be high,36.6 cycles/sec which is high compared to the rotational speed of the rotor of 1 cycle/sec.

2.5.3. Failure mode

In case of blade failure the structure has been calculate to withstand safely the resulting excess loads; figure 11 shows the load conditions for this failure mode.

2.6.Site selection-wind characterestics

The site which was selected for the installation of the wind generator is located 3 KM south of the town of Skyros, capital of the Skyros island. It is by the sea and the measurements of the last two years gave annual wind velocity at 10 meters height of 6.5 m/s. The site, even though better sites could be found on the island as regards wind energy potential, was selected on the basis

a. mean annual wind speed above 6.0 m/s.

b. site easily accessible for transportation and installation of the machine.

c. corrosion problems due to sand or salt should be exaggerated for research purposes.

d. the site should be easily accessible to locals and to tourists.

3.CONCLUSIONS

At this moment of reporting the full package of detailed drawings of the OA/100 KW HAWG has been given to the Greek Aerospace Industry and a proposal is awaited. Also preliminary market survey has been conducted in Greece of potential subsystem manufacturers. The component selection process has been finalised and firm economic and dilivery offers of sybsystems manufactures mainly in EEC have been received; if agreement is reached with the Greek Aerospace Industry on their economic and technical claims for constructing the parts, assembling the unit and erecting it on site, the OA/100 KW HAWG is envisaged for operation in 18 months time.

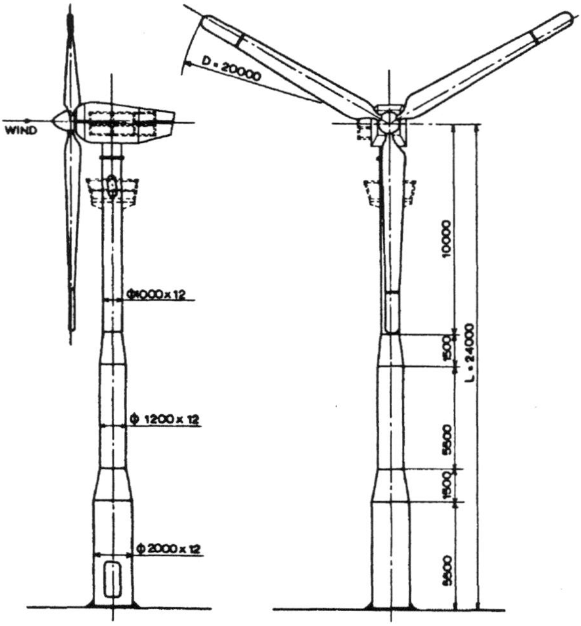

Fig.1: Wind Generator OA/100 KW

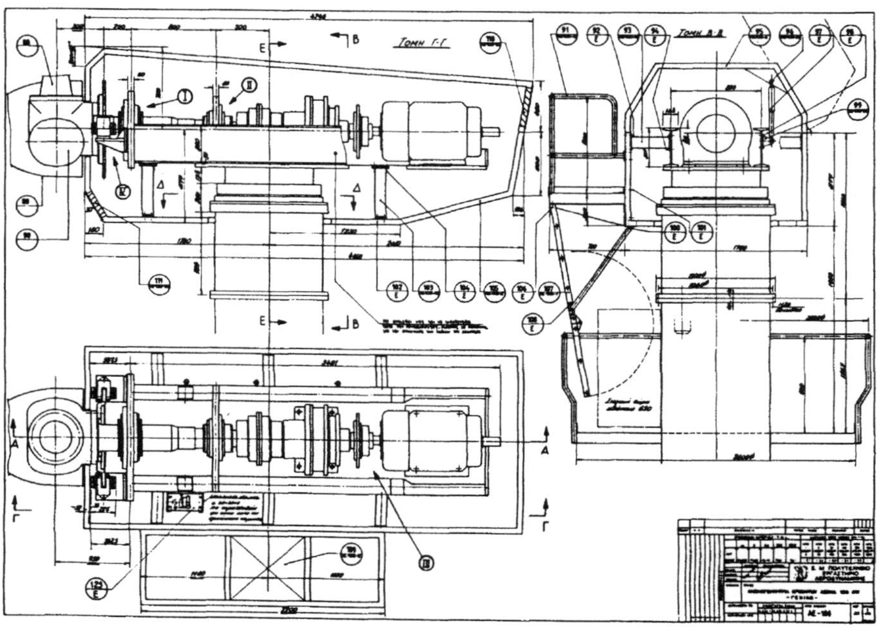

Fig. 2: The Nacelle

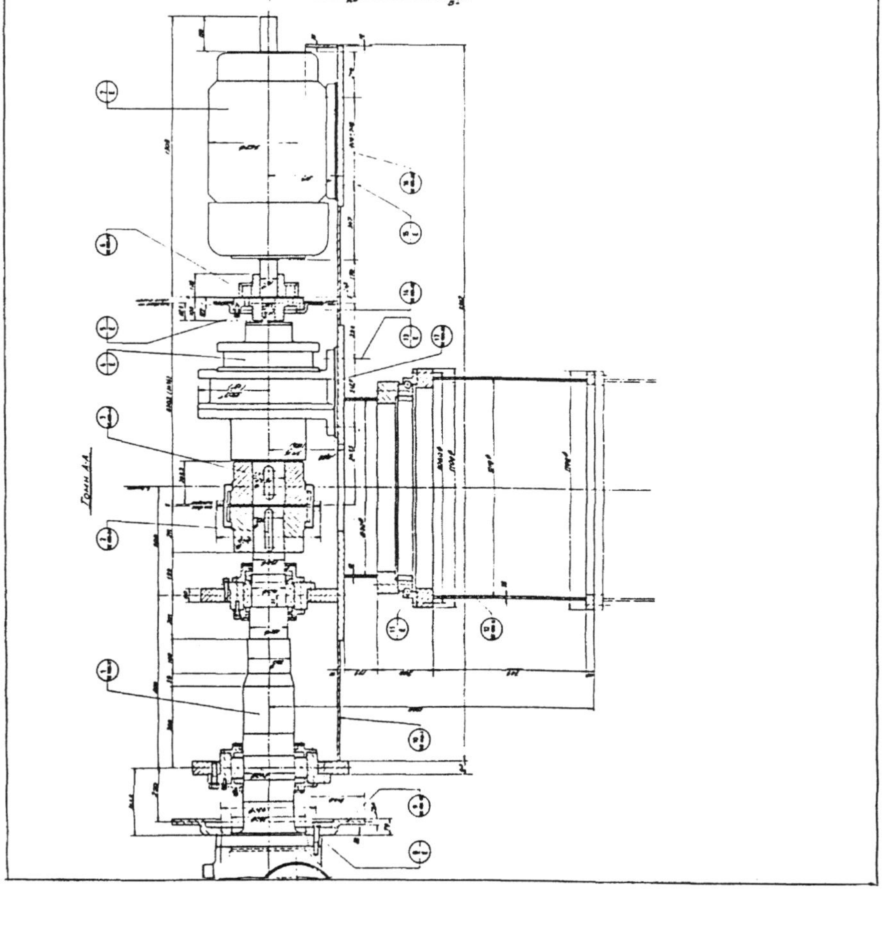

Fig. 3: The Nacelle

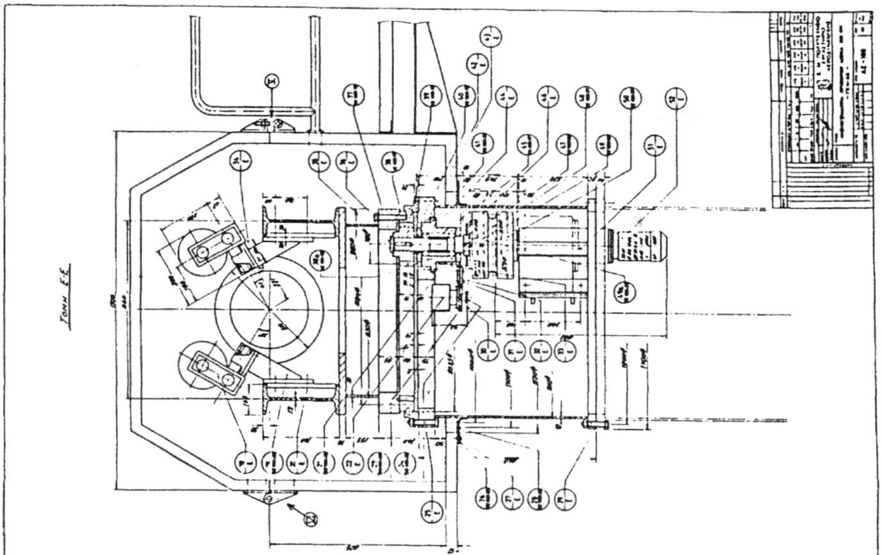

Fig. 4: The Drive train

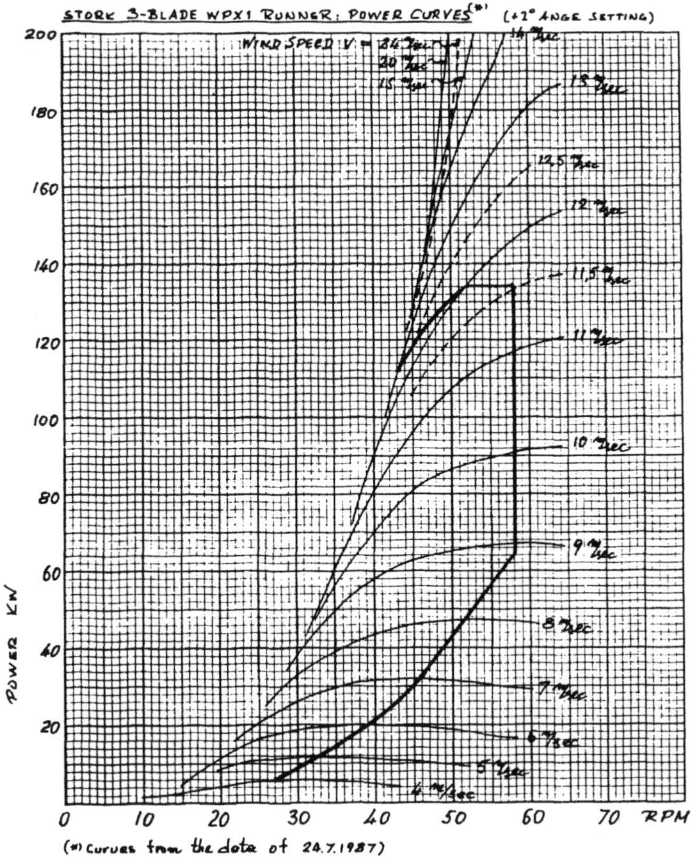

Fig. 5:Rotor Power WS RPM and wind speed

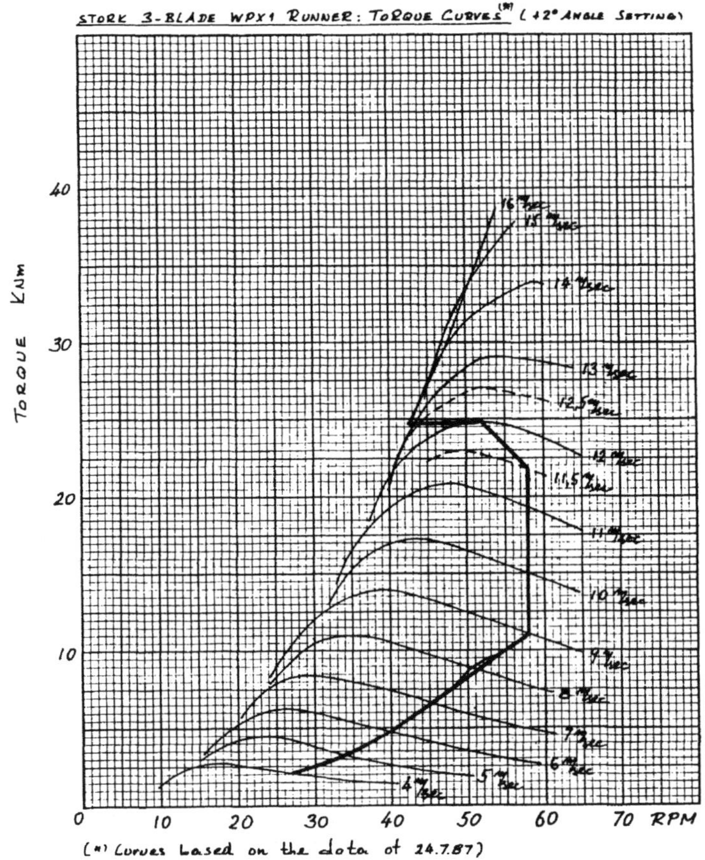

Fig. 6: Rotor Torque WS RPM and wind speed

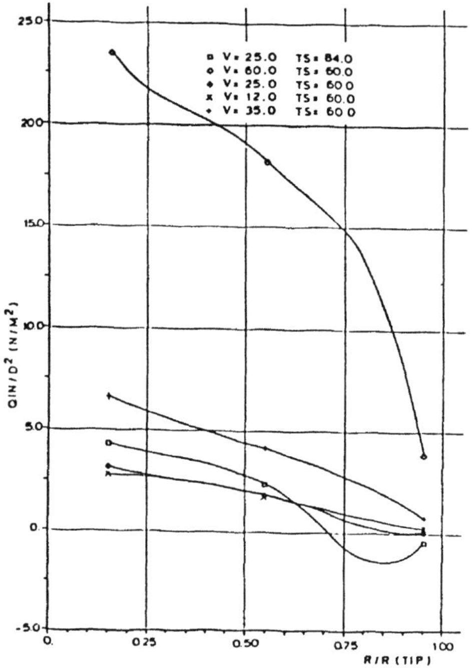

Fig.7: In plane stresses along the blade

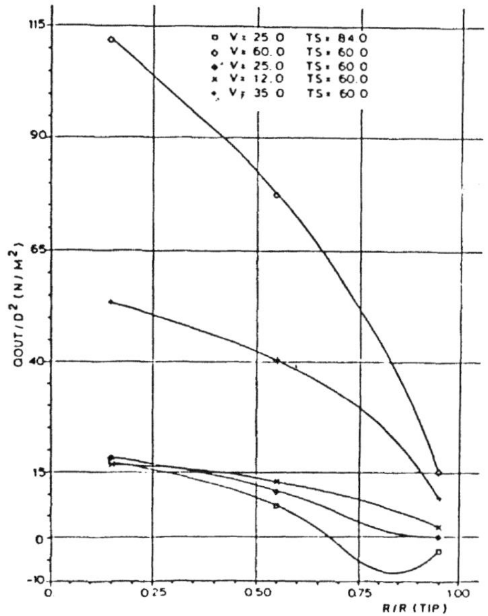

Fig.8:Out of plane stresses along the blade

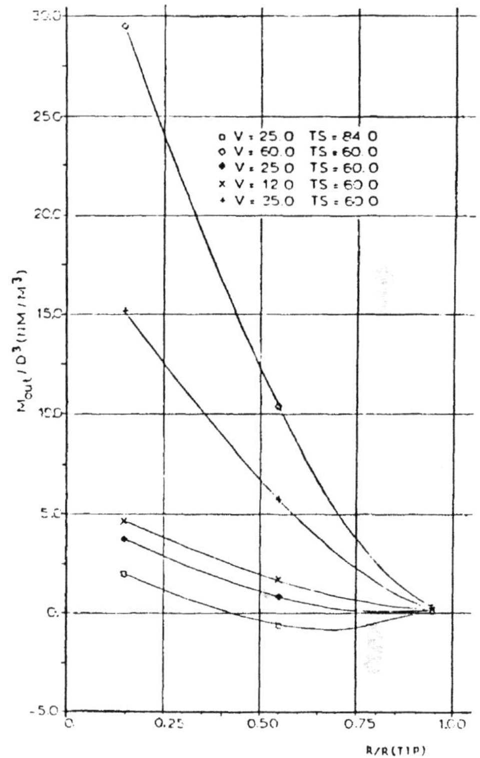

Fig.9:Out of plane Moment along the blade

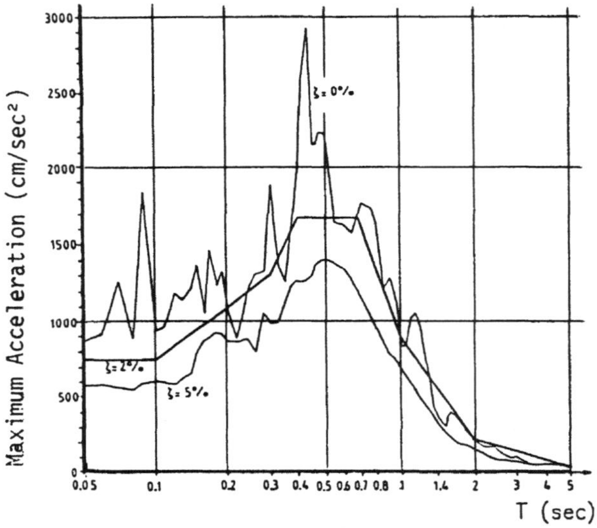

Fig.10:Spectrum of Greek Earthquaqes

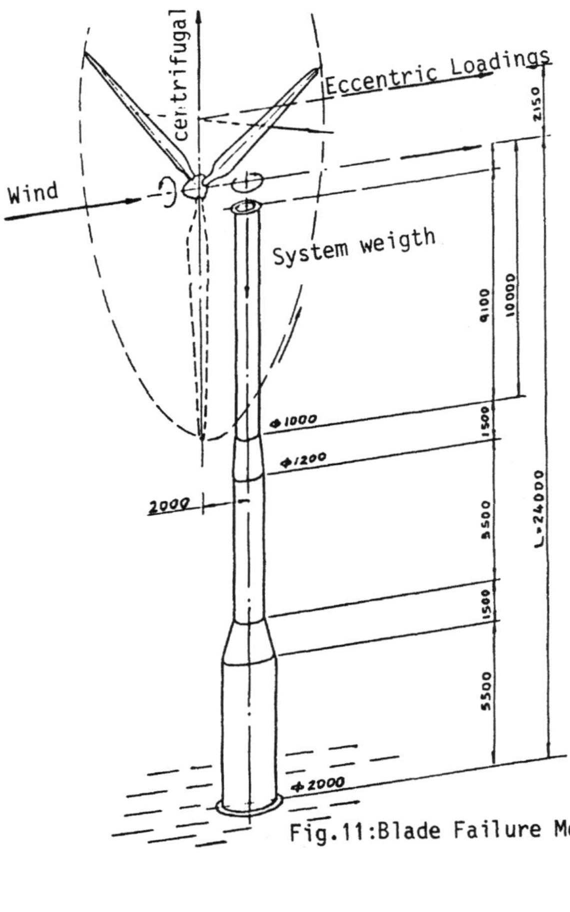

Fig.11:Blade Failure Mode

Project Nr 209/85 HE

A 100 KW DARRIEUS WIND TURBINE SYSTEM

G. Bergeles and N. Athanassiadis
Laboratory of Aerodymanics
Nat. Techn. University
Athens, Greece

SUMMARY

The report describes the aerodynamic design and the preliminary subsystem selection of a Darrieus vertical axis wind generator of rated power of 100 KW. The work has been undertaken by the laboratory of Aerodynamics of the National Technical University of Athens within a demonstration program sponsored by DGXVII and the Greek public power corporation. The wind generator has a height of 20 meters and a diameter of 18 meters and its two bladed troposkien shape blades are made of a NACA 0012 airfoil. The machine is expected to operate in the island of Skyros at annual average wind speeds of 6.5 m/s and diliver 180000 KWHs per year. At the time of reporting the Aerodynamic design of the machine has been completed as also the preliminary subsystem selection; as this design draws heavily on our experience of the design and construction of a horizontal axis wind generator of 120 KW it is expected that construction of the machine will finish within 18 months or so. The preliminary subsystem costs indicate that the cost per installed KW of the this machine will be much smaller than the cost projected for the HAWG OA/ 100 KW of similar size.

1.INTRODUCTION

The utilisation of wind energy has recently attracted much interest in the search for new energy sources; in contrast to the conventional wind the modern wind systems of wind energy utilisation are characterised by advanced technology. Since 1970 the main competetor to the horizontal axis wind generator (HAWG), the Vertical axis wind generator (VAWG), Darrieus type with curved blades, seems to stand on equal chances.Among the advantages of the VAWG compared to the HAWG are its symmetry to the wind which makes the yaw control system redendent, the special shape of the power coefficient curve that makes power control not necessary and the fact that the drive train system and the generator are located on the ground thus minimising maintainance and tower costs; in terms of overall system performance, the Darrieus vertical axis turbine seems to be competitive with horizontal axis wind generators; a performance penalty of approximately 7% is incurred generally by the Darrieus which is offset by its ability to operate without aerodynamic controls.
The design of the most cost effective wind turbine is a very complex process; the design undertaken in the present study aims at proving the reliability and cost effectiveness of a vertical axis wind generator of rated power of 100 KW a size which is optimum for the wind speeds and electricity grids of the Greek islands. The VAWG is selected to be installed on the island of Skyros next to the site of a Horizontal axis wind generator of 120 KW rated power, the OA/100 KW unit, as comparison of performance, of construction and maintenance will be possible for the two machines operating under similar wind conditions.

2.AERODYNAMIC DESIGN

The Aerodynamic design of the VAWG has been conducted using a computer code developed in the laboratory of Aerodynamics of the Technical University of Athens. The code employs the multiple streamtube theory of Strickland and its theoretical principles are shown in figure 1; the wind velocity well ahead the rotor is V, it reduces to (1-a)*V at the rotor area and is (1-2a)*V downstream of the rotor, where a is the axial reduction coefficient. The streamtube that passes the rotor is divided in many multiple streamtubes and on every streamtube the conservation of momentum is applied; the resulting integral force on each control volume is related to the forces acting on the blade as it twice crosses the elementary stream tube. The program requires airfoil data for the lift and drag coefficients as various angles of attack and Reynolds numbers. Figure 2 shows in graph form the data bank for the aerodynamic coefficients of the airfoil NACA 0012 used for the present calculations. A parametric study using the computer code indicated that a ratio of height to rotor diameter of around unity gives optimum results as regards maximum value of power coefficient and range of tip speed ratios; another parametric study of the effect of solidity on the power coefficient indicated an optimum range of solidities from .20 to .10; as regards distribution of blade chord the parametric study indicated that constant chord blades are maybe superior than variable chord blades. For given solidity of 0.15 a parametric study was conducted as regards maximum energy capture over the year for various rotor rotational speeds. The results indicated an optimum speed of 55 RPM. Finally the parametric study indicated high sensitivity of the results on the airfoil characterestic data as they are affected by changes in the Reynolds number. Figures 3 and 4 indicate the energy capture per year at the selected site at Skyros of 6.5 m/s mean annual wind speed and power coeffient of the finally selected machine; at this stage the typical geometrical characteristics of the VAWG are given in the following table.

PRINCIPAL CHARACTERISTICS OF THE VAWG

ROTOR HEIGHT	20 m
ROTOR DIAMETER	18 m
CHORD	0.6 m
AIRFOIL TYPE	NACA 0012
ROTATIONAL SPEED	55 RPM
NUMBER OF BLADES	2
SOLIDITY	0.13
MAXIMUM POWER COEFFICIENT	0.32
MAXIMUM POWER at 20 m/s	115 KW
CUT OUT WIND SPEED	25 m/s
SURVIVAL WIND SPEED	50 m/s
GENERATOR	Asychronous 100 KWel
ANNUAL ENERGY CAPTURE	180000 KWHs

3.EXTREME LOADS

The wind generator was designed to the following extreme wind conditions.

a. Nominal RPM with a wind speed of 50 m/s
b. Loss of electric power at 25 m/s, rotor overspeed to 75 RPM

c. Wind speed of 60 m/s at very low rotor RPM
Due to the time dependent character of the velocity field as the rotor rotates the loads on the blades fluctuate between a minimum an a maximum value; for the particular choice of the VAWG the blades at blade positions of 40,70 and 130 degrees azimuthal angle seem to be at the worst loading positions; figures 5 and 6 indicate the normal to blade forces along the length of the blade for the first two extreme load conditions; at 40 degrees azimuthal angle the parts of the blade at maximum diameter seem to be stressed at load condition of rotor overspeed, whilst at 70 degrees blade position the top and bottom parts of the blade seem to be stressed maximum at nominal RPM but with a gust of 50 m/s; at 130 degrees the middle part of the blades seem to have maximum loading. As regards chordwise force on the blade the critical condition is at 40 degrees azimuthal angle and at a loading condition of rotor overspeed, figure 7.

4. SUBSYSTEM SELECTION

The VAWG consists of a vertical axis of 24.5 metres height; at the top of the axis and at 4.5 meters from ground (height 20 meters) are fixed via two flanges the two blades of the wind generator. The drive train mechanism, the generator and the supporting subsystems are housed in a small building on the ground at the base of the shaft. A general layout of the wind generator with preliminary dimensioning and component selection is shown in figure 8.A logical approach to the design of an optimised VAWG system is to select the commercially available subsystems and define the remaining components accordingly. Particular details of the selected subsystems are given in the following.

4.1. Rotor design

The rotor consists of two extruded aluminium blades, each having a chord of 60 cm and a NACA 0012 airfoil section; the two blades are attached to the hollow tube via flanges. The rotor diameter is 18 meters. Each blade is formed by three parts into a similar to troposkien shape; the three sections are bolted together using inserts. Provisions have been taken of installing braces from the blade junctions to the shaft, if needed. The torque tube is assembled from four sections; at the top of the shaft a bearing assembly housed in flanged type construction takes the loads imposed by the possible differences in pretension and of the vertical thrust of the guy wire system. The lower bearing assembly balances the aerodynamic loads and the weights of the rotor and transfers them to the steel structure.

4.2. Brake system

Two independent brake sets are incorporated on the rotor low speed shaft and on the high speed shaft. The low speed shaft brake is a fail safe hydraulically actuated brake of nominal diameter of 910 mm; the high speed shaft disc brake is an electric fail safe which is designed to operate for parking the rotor and also in the event of malfunction of the low speed hydraulic brake. Any brake system is selected to be able to stop the rotor in overspeed conditions and in wind speed of 50 m/s.

4.3. Controls

The operation and the surveyance of the system is performed by the wind turbine's microprocessor. The microprocessor is a programmable logic controller of industrial type where the algorithm of start up, operation, stop at cut out speed and monitor of special critical parameters have been hardware preprogrammed. The output signals act mainly on the rotor brakes. A personal computer equipped with an A/D converter independently monitors the performance of the wind generator.

5.SITE SELECTION

The VAWG will be installed on the island of Skyros at a site close to the one already selected for the installation of a HAWG of 120 KW power. The site as wind velocity measurements show for the last two years has a mean annual wind speed at 10 m height of 6.5 m/s and a predominant wind direction NE coming from the sea; the site was selected apart from reasons of high cost of electricity production which prevail all over the island and for reasons of

a. easy accessability to the site for transportation and installation of the machine;
b. corossion problems due to salt and sand typical to be found in future constructions in Greek islands should be studied;
c. touristic attraction for diffusion of the concept of alternative energy sources;

6.CONCLUSIONS

At this stage of reporting the preliminary aerodynamic design has completed and the final design and subsystem components is well underway;since this design draws heavily on our experience from the earlier design of the HAWG, OA/100 KW unit, of similar size the construction of this machine will be completed almost together with that of the HAWG even though started a year later and therefore if agreement is reached soon with the subcontractors the machine will be in operation within 18 months.

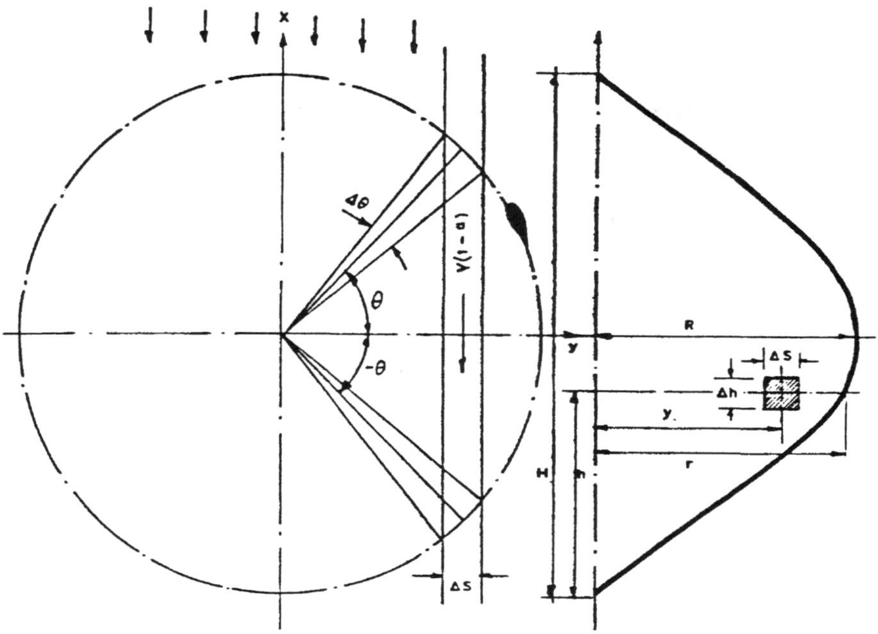

Fig.1: The multiple Streamtube theory

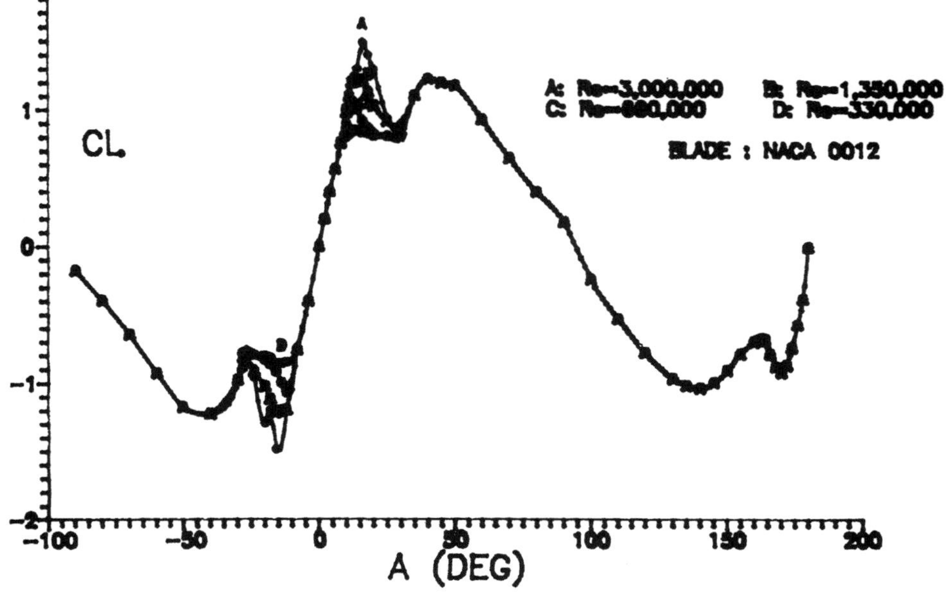

Fig.2: Aerodynamic airfoil data

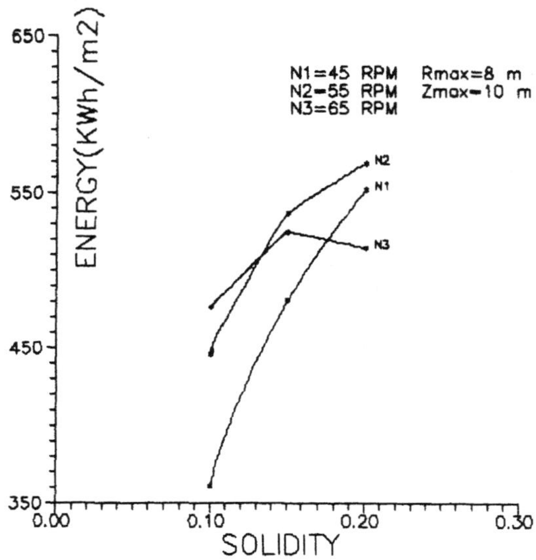

Fig.3: Optimisation on energy capture

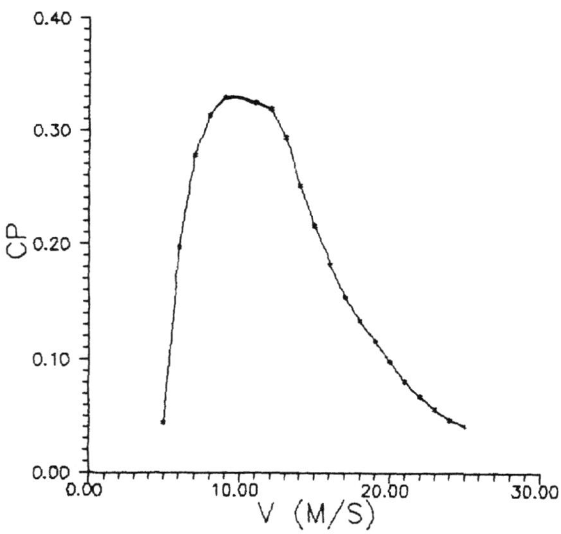

Fig.4: Power factor of the VAWG

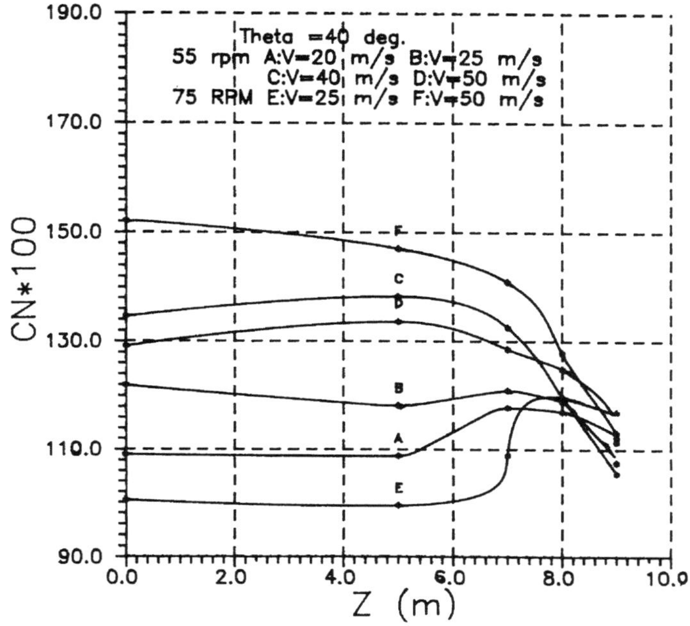

Fig.5: Normal forces along the blade, $\varphi=40°$

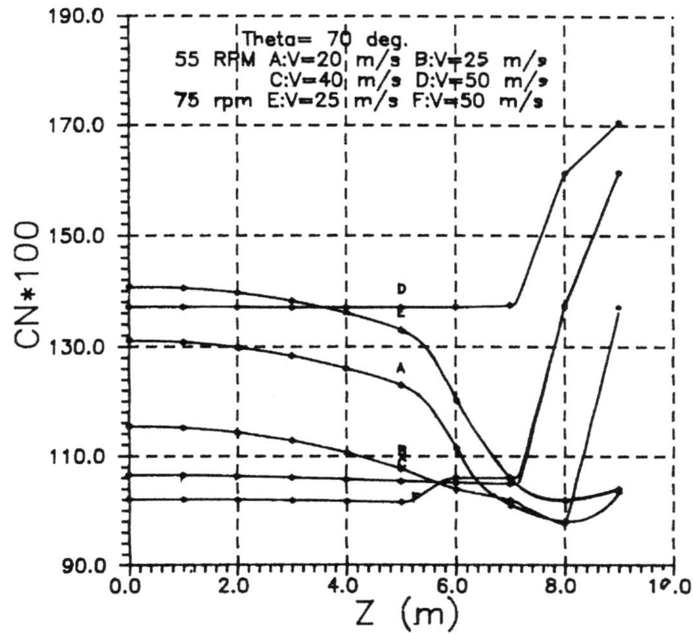

Fig.6: Normal forces along the blade, $\varphi=70°$

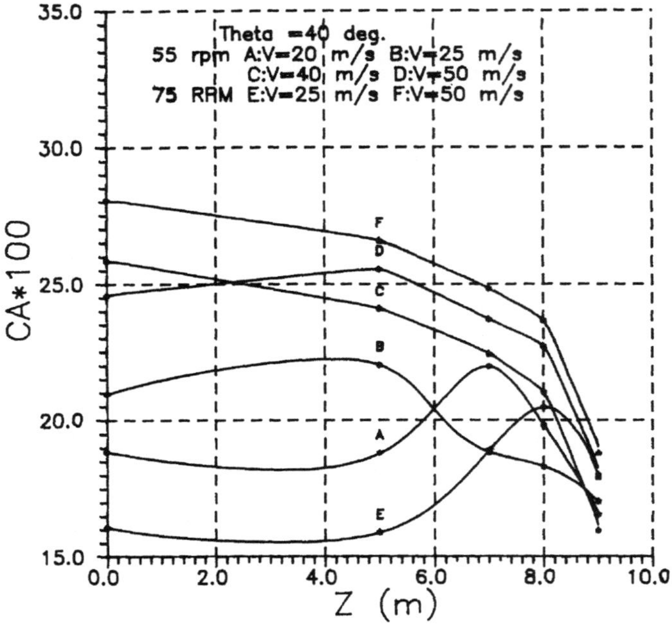

Fig.7: Axial forces along the blade, φ=40°

Fig. 8

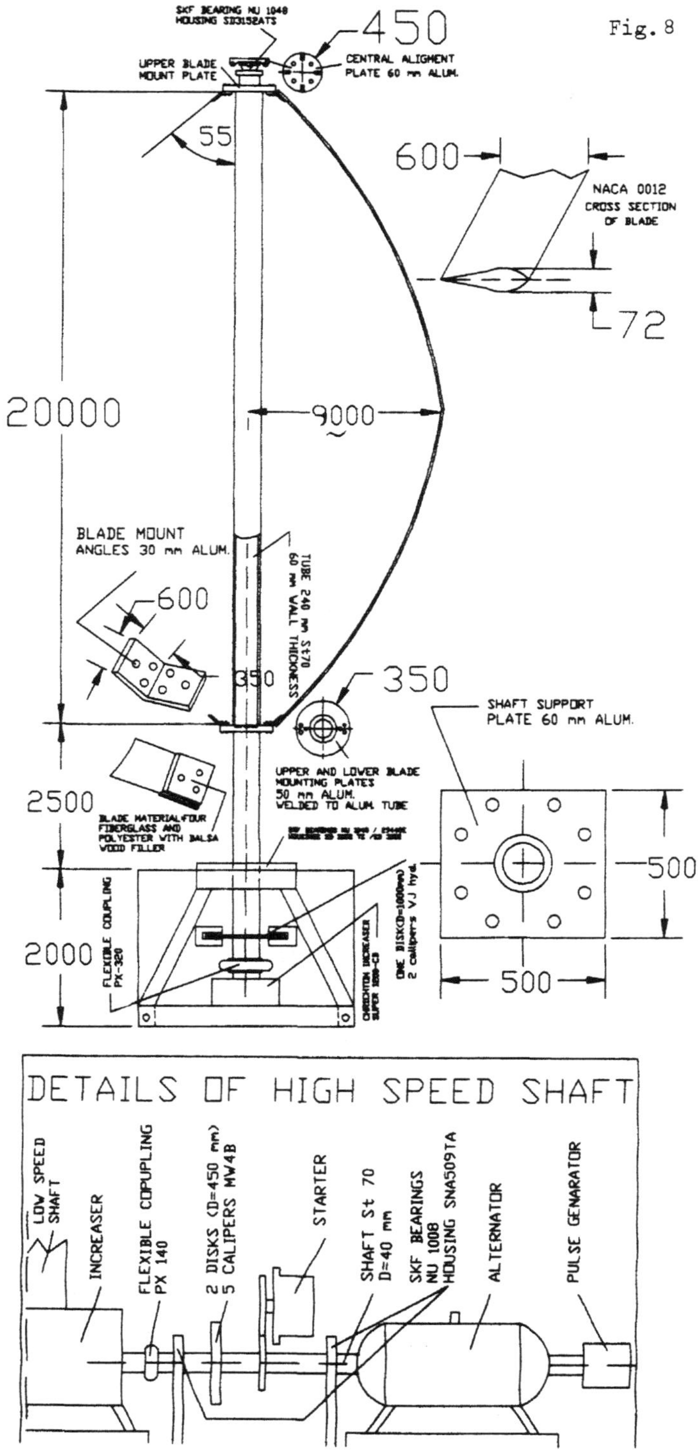
SKF BEARING NU 1048
HOUSING SD3152ATS
UPPER BLADE MOUNT PLATE
450
CENTRAL ALIGMENT PLATE 60 mm ALUM.
55
600
NACA 0012 CROSS SECTION OF BLADE
72
20000
9000
BLADE MOUNT ANGLES 30 mm ALUM.
600
350
TUBE 240 mm St70 60 mm WALL THICKNESS
350
SHAFT SUPPORT PLATE 60 mm ALUM.
UPPER AND LOWER BLADE MOUNTING PLATES 50 mm ALUM. WELDED TO ALUM. TUBE
2500
BLADE MATERIAL FOUR FIBERGLASS AND POLYESTER WITH BALSA WOOD FILLER
500
500
FLEXIBLE COUPLING PX-320
2000
ONE DISK(D=1000mm) 2 calipers VJ hyd.
DETAILS OF HIGH SPEED SHAFT
LOW SPEED SHAFT
INCREASER
FLEXIBLE COPUPLING PX 140
2 DISKS (D=450 mm)
5 CALIPERS MW4B
STARTER
SHAFT St 70 D=40 mm
SKF BEARINGS NU 1008 HOUSING SNA509TA
ALTERNATOR
PULSE GENARATOR

Projects Nr 157/83 IT, 147/85 IT

TWO, SMALL AND MEDIUM POWER RATED, AUTONOMOUS WIND-DIESEL SYSTEMS

A. BLOTTO FINADRI, C. PALMARI, M. ROTONDI
AERITALIA - Società Aerospaziale Italiana

Summary
The paper describes two autonomous Wind-Diesel systems, the first (small power rated) is already installed and under testing, the second (medium power rated) is presently in the designing stage.
For the small power system we will describe technical details, synchronization and parallel operation of synchronous generators, Diesel sets control logic and inverters operation.
For the medium power system, general description, designing criteria and operation modes will be given.

1. INTRODUCTION

The small power system uses AIT 03 Wind Turbine developed by Aeritalia mainly for small completely isolated system to feed battery charger rectifier (1) without any parallel operation with the back-up Diesel provided for emergency power.

Table I furnishes technical data and characteristics for AIT 03 Wind Turbine.

The standard construction of AIT 03 WT is equipped with three phase brushless synchronous generator.

The small power system (named CADA-POSH) is located in Calabria (near Villa S. Giovanni).

The main objectives of this demonstration plant are explained at the end of Chapter 2.

Synchronization through WT speed regulator, Diesel sets control logic, inverters operation are the main innovative features (see Chapter 3 and 4).

The medium power system uses AIT 02 (MEDIT) Wind Turbine developed by Aeritalia mainly for parallel operation with external grid (2).
Table II furnishes technical data and characteristics for AIT 02 (MEDIT) Wind Turbine.

The standard construction of MEDIT is equipped with asynchronous generator (squirrel cage-induction).

The medium power system will be located in Abruzzo (Montorio al Vomano).

The main objectives of this demonstration plant are explained at the end of Chapter 5.

2. SMALL POWER SYSTEM DESCRIPTION

The small power system (see Fig. 1) includes mainly:
1 AIT 03 WT equipped with three phase brushless synchronous generator rated 20 kVA, 380 V, 50 Hz, 1500 RPM;
2 Diesel Generator Sets each equipped with three phase brushless synchronous generator rated 20 kVA, 380 V, 50 Hz, 1500 RPM;
1 Forced Commutated Inverter, 380 V - 50 Hz, three phase, rated 8 kW;
1 Battery Charger Rectifier/Line Commutated Inverter, 380 V - 50 Hz three

phase, rated 8 kW;

1 Lead Acid Storage Battery 120 V, 100 Ah.

Synchronous generators for Wind Turbine and Diesel sets are identical and are equipped with incorporated solid state voltage regulators, compound type with over excitation in order to substain short circuit currents and with voltage drooping characteristic for reactive power balance.

Diesel engines are two cylinders, four-stroke type, air cooled, VM SUN 2105 type equipped with Barber-Colman (Plus 1) solid state electronic speed governor and with Basler PRS/370 auto-synchronizer.

This small power Wind/Diesel system named CADA-POSH was already described in EWEC Conference 1986 (3) with particular regard to operation modes and computer simulations.

This plant was designed and built using standard and proved components as much as possible.

Since synchronous generator was used for WT, we don't need electromagnetic clutch to leave Diesel set generator running as synchronous compensator when the engine is stopped (as in the case of asynchronous generator for WT), that means we can adopt truly standard Diesel generator set and we can build this system where Diesel generator set is existing.

In our case we have two Diesel generator sets, one is considered the first (N. 1) to be started by the logic in automatic operation, the second (N. 2) will be started if N. 1 fails or in the case of exceptional overloads.

With a selector switch on the control panel we can interchange their role (starting priority).

The Diesel engine governor is adjusted with a speed drop of 5% (in order to have good active power balance), that means in our case that the frequency is 50,8 Hz no load and 48,5 Hz full load (See Fig. 2).

The main objectives of this demonstration plant are the testing of:

- synchronisation and parallel operation of WT synchronous generator with Diesel set generator
- inverters operation and control logic in order to avoid too frequent starting and low load operation of Diesel engines
- continuity of power supply

3. GENERATORS SYNCHRONISATION. WT SPEED REGULATOR

The Basler auto-synchronizer (acting on diesel engine speed governor) is used when one diesel set is started, while the a.c. common bus are already fed by another Diesel set, or WT or forced commutated inverter.

The synchronisation of WT with A.C. common bus already fed by one Diesel set, could be done using Basler auto-synchronizer of the Diesel set.

We preferred to equip also the WT with speed regulator and auto-synchronizer custom built.

Since AIT 03 WT is fixed pitch type, the best way to control the speed in isolated (stand-alone) operation is the use of dump resistive load with electronic control.

In this case we adopt a fully controlled six pulse thyristor bridge feeding a fixed value resistor (active power dissipated is proportional to

$$\frac{V^2}{R} = \frac{V^2_{max} \cdot \cos^2\alpha}{R}$$

This kind of dump load is very simple but has the disadvantage to draw reactive power from the generator and to generate harmonic currents that produce voltage wave shape distorsion.

The closed loop speed regulator (See Fig. 3) compares speed feedback, from tachometer generator coupled to WT, with speed reference signal.

The speed error is amplified and acts on thyristor bridge firing circuit in order to vary electrical torque.

The WT auto-synchronizer acts on the speed regulator trying to obtain the WT generator frequency as much as possible equal to AC common bus frequency (regulator speed reference is taken from the bus).

When the two frequency are practically the same, the phase angle circuit comes in action and when the phase angle (between the two generators) remains within a preselected band width for a certain time, closure signal to WT generator contactor is given.

For safety reason we perform a two-steps contactor closure (first step through a damping resistor).

When the synchronous operation is finished, the speed reference is raised to 50,8 Hz (no load value for Diesel set).

When WT active power is greater than consumers loads, the excess power tries to overspeed the Diesel engine.

The speed governor regulating loop becomes open.

To limit the overspeed, the speed regulator acting on dump load comes in action, limiting the frequency to roughly 52 Hz (See Fig. 2).

After a certain time of low load operation, Diesel set is stopped, synchronous generator disconnected from the A.C. bus bars, while WT remains alone, in isolated operation with the speed (frequency) controlled by the speed regulator acting on dump load.

4. DIESEL SETS CONTROL LOGIC, INVERTERS OPERATION

Diesel set N. 1 is automatically started when battery voltage is low or when line-commutated inverter is operating for more than 5 minutes.

Diesel set N. 2 is automatically started if Diesel set N. 1 starting fails or when line-commutated inverter is operating for more than 5 minutes.

Diesel N. 1 is stopped to avoid low load operation when the active power supplied is less than a preselected value (10%) for a certain time, provided battery voltage is sufficiently high.

Diesel N. 2 is stopped when the active power supplied from the two Diesel sets is less than 70% of one Diesel set rated power for a certain time.

The line commutated inverter can be connected to A.C. common bus only if at least one synchronous generator is already in operation.

The connection operation is made through a frequency relay which senses overload condition (See Fig. 2).

The inverter operation is "symmetrical" to dump load. Dump load control avoids the raise of frequency, inverter operation avoids too much drop.

The inverter action is considered a "help" for the WT and/or a Diesel generators.

The inverter reduces the numbers of starts of the Diesel engines, since the consumers overloads and wind speed decreases (gusts) are often temporary (less of 5 minutes duration).

In addition to line-commutated inverter we have in this case also a forced-commutated inverter which is dedicated to feed very low consumers loads (for istance lighting during night) avoiding to keep in operation one Diesel set, in case of wind absence.

This kind of converter can be connected to A.C. common bus only if all synchronous generators are disconnected.

The forced-commutated inverter remains in operation until battery voltage drops down.

When this happens, Diesel set is started feeding the consumers loads and charging the battery.

5. MEDIUM POWER RATED SYSTEM DESCRIPTION

The medium power rated system (See Fig. 4) includes mainly:

1 AIT 02 (MEDIT) WT equipped with three phase asynchronous generator (squirrel cage-induction) rated 280 kVA (250 kW) 380 V, 50 Hz, 1515 RPM;

1 Diesel Generator Set equipped with three phase brushless synchronous generator rated 375 kVA, 380 V. 50 Hz, 1500 RPM and with electromagnetic clutch in order to leave generator running as synchronous compensator while Diesel engine is at standstill.

In this case we choose asynchronous generator for WT, since it is the standard solution for future connection to external grid.

The synchronous generator is equipped with incorporated solid state voltage regulator, compound type with overexcitation in order to substain short circuit currents.

Battery storage is not provided, since the isolated operation (stand-alone with Diesel engine at standstill) is intended only for consumers loads for which it is possible to adopt load management (shedding) technique.

The main objectives of this plant are the testing of:

- electrical transient (inrush current and voltage drop) due to asynchronous generator connection with synchronous
- frequency control during isolated (stand-alone) operation (dump load versus pitch control).

6. MEDIUM POWER SYSTEM OPERATION MODES

Three operation modes are possible:

- Diesel set stand-alone
- Diesel set in parallel with WT
- WT stand-alone

In any case, operation begins starting the Diesel set which is completely controlled by engine speed governor and generator voltage regulator.

If wind speed is above cut-in and under cut-out value, we can start WT acting on pitch control.

The acceleration is controlled through a speed regulation loop and we bring the asynchronous generator speed as much as possible equal to the synchronous speed corresponding to Diesel generator frequency.

When this condition is reached, we close the WT generator contactor and we will have an electrical transient lasting few cycles.

After the connection between the two generators is made, we are in the following situation:

- Diesel synchronous generator keeps AC bus bars voltage constant and supplies reactive power to asynchronous generator and external consumers
- Diesel engine governor keeps speed (frequency) constant until active power supplied from WT is less than active power requested from consumers, or until a minimum active power is supplied from the Diesel set.

If active power supplied from WT becomes equal or greater than consumers loads, speed governor regulating loop becomes open and we have the risk of overspeed.

In this plant we intend to avoid overspeeding of medium power Diesel engine and we wish to guarantee a minimum load for the Diesel.

We can obtain that reducing the active power supplied from WT acting on pitch control or using electronic controlled dump load.

WT stand-alone operation is possible when wind power is greater than consumers loads.

Since the wind power is fluctuating, the simplest way to get WT stand-alone operation is to vary the consumers loads in proportion to available wind power.

This can be done for istance if consumers loads is made with several steps of different size, easy to connect and disconnect, (heating resistors, pump motors etc.).

During WT stand-alone operation we can open electromagnetic clutch and stop the Diesel engine.

REFERENCES

(1) V. Fisauli, C. Palmari - Wind System AIGE-AIT 03 for Stand-Alone Users: Configuration and Experimental Results. EWEC 1986 Rome - E31

(2) S. Avolio, G. Gaudiosi, F. Iacovoni, G. Sorli - MEDIT: the Forerunner of Medium Size WTG Family - EWEC 1986 - Rome D2

(3) O. Honorati, F. Crescimbini, G. Sorli, C. Palmari - Wind Diesel Systems: Problematics and Prospects EWEC 1986 - Rome E16.

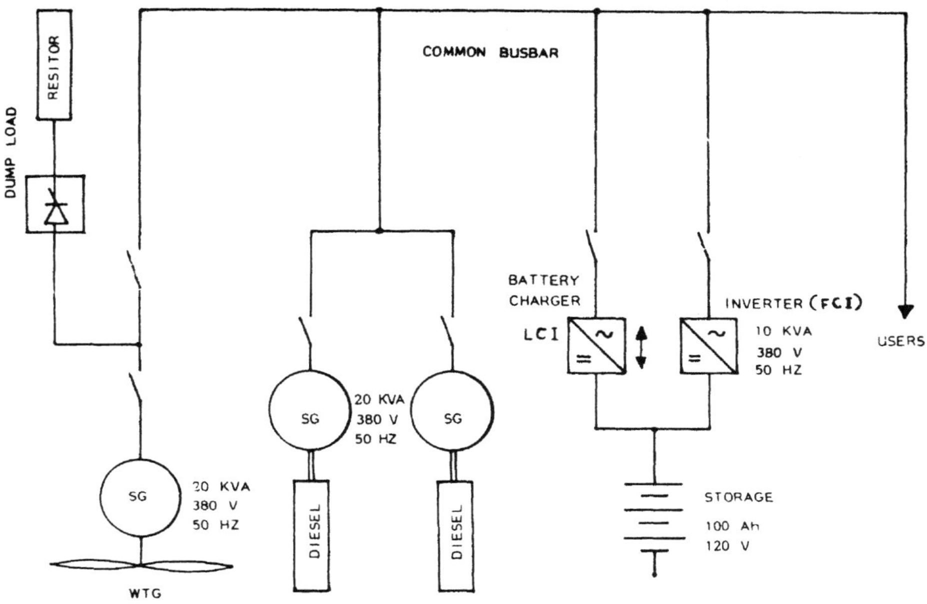

FIG. 1 : SMALL POWER SYSTEM ELEMENTARY DIAGRAM

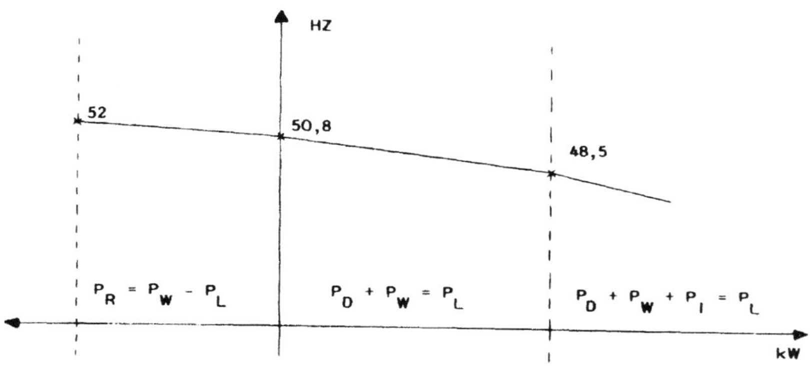

P_W = WT ACTIVE POWER

P_D = DIESEL SET ACTIVE POWER

P_I = LCI ACTIVE POWER

P_R = DUMP RESISTOR POWER

P_L = CONSUMERS LOAD

FIG. 2 : SMALL POWER SYSTEM kW/FREQUENCY CHARACTERISTICS

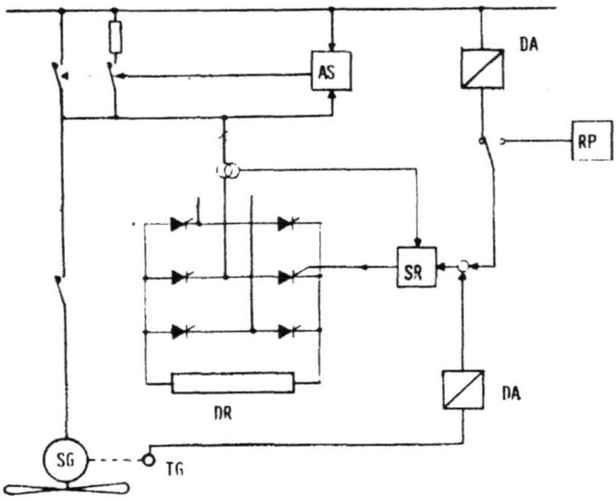

SG = SYNCHRONOUS GENERATOR
TG = DIGITAL TACHO-GENERATOR
DA = DIGITAL/ANALOG CONVERTER
RP = REFERENCE SIGNAL **AFTER** PARALLEL
SR = SPEED REGULATOR
DR = DUMP LOAD RESISTOR
AS = AUTOSYNCHRONIZER

FIG. 3 : SMALL POWER SYSTEM SPEED REGULATOR ELEMENTARY DIAGRAM

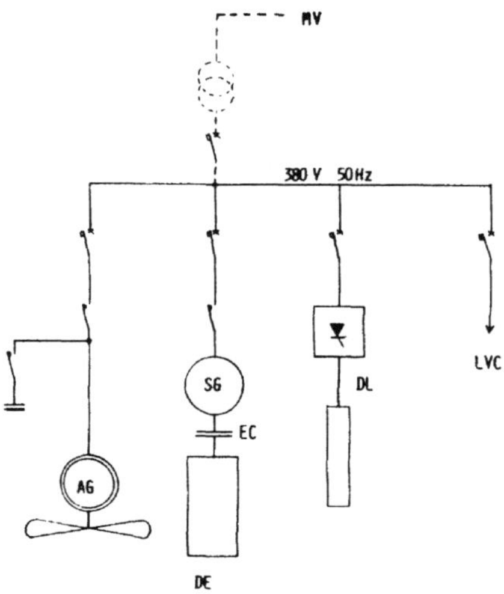

AG = ASYNCHRONOUS GENERATOR
SG = SYNCHRONOUS GENERATOR
EC = ELECTROMAGNETIC COUPLING
DE = DIESEL ENGINE
DL = DUMP LOAD
LVC = LOW VOLTAGE CONSUMERS
MV = FUTURE MEDIUM VOLTAGE CONNECTION

FIG.4 : MEDIUM POWER SYSTEM ELEMENTARY DIAGRAM

GENERAL CHARACTERISTICS		CONTROL SYSTEM	
Number of blades	2	Power regulation	Stall regulated
Rotor diameter	8,6 m		Fixed pitch
Rotor surface area	58 m^2		
Solidity	6%	Alignment type	passive
Rotor position	down wind		
Hub height	10 meters		
Rotation direction	clockwise (looking windwards)	OVERGEAR	
Rotation speed	93 rpm	Type	epicyclic
Coning angle	4°		
Tilt angle	0°	Gear ratio	16,16
Rated blade pitch setting	2°		
OPERATING CHARACTERISTICS		ELECTRIC GENERATOR	
Rated power	17 kW	Type	brushless synchronous
Wind speed at 10 m (a = 1/7)			
Cut in	5 m/s	Voltage	380 V, 50 Hz, 3 ø
Rated	20 m/s	Rated power	20 kVA
Cut out	20 m/s	Synchronous speed	1500 rpm
Survival (non-operative)	50 m/s	Efficiency	85%
ROTOR		TOWER	
Blade		Type	tubular
Material	Composite (GFRP)	Material	FE 42 B steel
Airfoil sections	NACA $63_3 6XX$	Diameter	0,457 m
		Height above ground	9,5 m
Chords	0.297 m (tip)	Foundations	reinforced concrete
	0.510 m (root)		
Taper	linear		
Twist	11.3° (linear)		
Hub			
Type	rigid		
Material	steel		

TABLE 1 - AIT 03 TECHNICAL DATA

GENERAL CHARACTERISTICS		CONTROL SYSTEM	
Number of blades	2	Operation and regulation	
Rotor diameter	32.00 m	Type	automatic
Rotor surface area	804 m^2	Central unit	microprocessor
Solidity	4.5%	Power regulation	
Rotor position	up wind	Type	blade pitch variation
Hub height	26 meters		
Rotation direction	anticlockwise (looking windwards)	Pitch actuator	hydraulic
		Actuation speed	6 degrees/s (max)
Rotation speed	40.24 rpm	Alignment	
Coning angle	4°	Type	active
Tilt angle	6°	Actuator	hydraulic
Rated blade pitch setting	0°	Yawing rate	2 degrees/s (max)
OPERATING CHARACTERISTICS		**OVERGEAR**	
Rated power	225 kW	Type	3-way
Wind speed at 10 m (a = 1/7)		Number of steps	2+1 bevel gear pair
Cut in	4.2 m/s	Gear ratio	37.27
Rated	10.1 m/s	Efficiency	95%
Cut out	20 m/s		
Survival (non-operative)	50 m/s		
		ELECTRIC GENERATOR	
ROTOR		Type	squirrel-cage induction
Blade		Voltage	380 V, 50 Hz, 3 Ø
		Rated power	250 kW
Material	Composite (GFRP)	Synchronous speed	1500 rpm
Airfoil sections	NACA 4412 (end) NACA 4420 (root)	Efficiency	94.8%
Chords	0.600 m (tip) 1.800 m (root)		
		TOWER	
Taper	linear		
Twist	8° (linear)	Type	three cylindrical segments
Hub		Material	COR-TEN B steel
		Diameter	1.6 m
Type	rigid	Height above ground	25.00 m
Material	steel	Foundations	reinforced concrete

TABLE 2 - AIT 02 (MEDIT)TECHNICAL DATA

Projects Nr 337/83 FR, 92/86 FR, 154/86 FR

DEVELOPMENT OF AUTONOMOUS WIND ENERGY POWER PLANTS

J.M. NOEL
AEROWATT-INTERNATIONAL
6, avenue des Coquelicots
94385 BONNEUIL Cédex (FRANCE)

Summary

Since 1983, the Aerowatt company and its successor, Aerowatt International, have presented a number of energy demonstration projects in response to calls for tender issued by the Commission of the European Communities. The majority of these systems deal with autonomous wind energy power plants delivering electricity in a standard industrial format. These projects are located in the GLENAN archipelago, situated off the southern coast of BRITTANY. Their purpose is provide electricity on various islands in this group. The purpose of this paper is to present the results already obtained (completed by information coming from similar installations not financed by the Commission) and to indicate the main axes for further development of this technology.

1. INTRODUCTION

The GLENAN archipelago, a part of the FOUESNANT municipality, is located off the southern coast of the Department of FINISTERE in the BRITTANY region of FRANCE. Four islands in this group (BANANEC, DRENEC, CIGOGNE, PENFRET) are used as bases for a sailing school owned and operated by the CENTRE NAUTIQUE DES GLENANS (CNG) which offers study programs centered around school holidays, including the Christmas period. The combined total of students can reach 400 during the month of August.

The CNG has a strong ecological orientation and the bases have, in the past, been without basic comfort features (drinking and sanitary water supplies pumped by hand, lighting and cooking assured by propane gas, etc.). No electricty generator sets were operated by the CNG.

One island in the GLENAN group, SAINT-NICOLAS, is inhabited by fishermen and a representative of the FOUESNANT municipality on a year-round basis. This island is equipped with a 32 kW generator set which is used to pump drinking water from a submersible pump into a cistern and then into an elevated storage tank and also to charge a 120 V battery bank used to supply DC power for lighting.

The development of tourism on SAINT-NICOLAS (there can be as many as 1500 daily visitors in the summer months) and the comfort of the students at the CNG base on this island are limited by the lack of drinking water. When the islands in the GLENAN group were only inhabited and cultivated by a permanent, indigenous population, the main source of drinking water was rainfall collected in cisterns. This resource is not sufficient to meet the needs of the large, non-permanent population which comes to the islands during the driest period of the year. These islands are too small to have a sufficient, sub-surface water supply.

In conclusion, these islands can be good bases to demonstrate the efficiency of decentralized, autonomous wind energy power plants used to provide energy for a number of daily requirements.

2. PROJECTS IN PROGRESS

Currently, four projects, financed in part by the Commission of the European Communities, are in progress in the GLENAN archipelago:

Contract WE/337/83/FR : Autonomous wind energy power plant on PENFRET island (CNG) supplying power for general purposes (lighting, refrigeration, small tools, pumping); the fresh water supply on PENFRET is greater than on other islands in the GLENAN group.

: Autonomous wind energy power plant on BANANEC island (CNG) used primarily to desalinate sea water for fresh water requirements.

Contract WE/154/86/FR : Autonomous wind energy power plant on SAINT-NICOLAS island used primarily to desalinate sea water and secondarily to meet the electricity needs of several community installations.

Contract WE/092/86/FR : Autonomous wind energy power plant on DRENEC island (CNG) used primarily to desalinate sea water.

These projects are in the following stages of completion:

PENFRET

The system was installed in the fall of 1986; it was fully operational during the 1987 training sessions and during the Easter sessions of 1988. It will be further used during the summer of 1988 and in the years to come.

BANANEC

This system, whose installation has been delayed for a number of reasons, including a recent environmental objection raised by the MINISTERE DE L'EQUIPEMENT ET DU LOGEMENT, should be operational before the end of summer 1988.

SAINT-NICOLAS

The energy segment of this system was installed and commissioned in April 1988 after being delayed due to poor weather conditions during the fall of 1988. The associated desalination equipment will be operational in July 1988.

DRENEC

An environmental consideration remains as an obstacle to the completion of this project. The regional office of the AGENCE FRANCAISE POUR LA MAITRISE DE L'ENERGIE is directing a comprehensive action to obtain a general authorisation for the installation of wind energy systems in protected areas.

3. DESCRIPTION OF AUTONOMOUS WIND ENERGY POWER PLANTS

Each of the above systems is founded on a similar principle:

- the run-of-the-wind AC energy output of the windgenerator is transformed into DC by a rectifier,
- the DC current is transformed into industrial format AC current by an inverter,
- as a function of the relation between the DC current available at the rectifier outputs and the DC current demanded by the inverter, all or part of the available current is stored in or drawn from the battery,
- the AC current is distributed normally as long as the charge level of the battery remains above a predetermined level.

The technical differences between the various projects are primarily the result of continuing efforts to lower the production cost of the energy being distributed. This evolution is centered around two elements of the system:

- the windgenerators,
- the dual-conversion (AC/DC/AC) equipment.

3.1. Windgenerators

The first three projects, PENFRET, BANANEC and SAINT-NICOLAS are or will be equipped with similar windgenerators from the WM 70 series.

This machine is equipped with a rapid, two-blade rotor (tip speed: about 75 m/s) operating upwind of the tower. Orientation is assured by a tail vane. The rotor drives a self-excited, asynchronous generator of 6 or 12 kW rated power through a gearbox using parallel helicoïdal gears with step-up ratios of 1:8 (6 kW machine) or 1:7 (12kW machine).

The rotor diameter is 7 meters; the power and energy yield characteristics are given in the Appendix. Aerowatt International has shipped 120 machines of this type around the world and considers this equipment to have reached technical maturity.

The remaining project, DRENEC, will be equipped with a UM 100(Vms40)/10.000/BAT(50)NC windgenerator with a rated power of 12 kW; this machine is equipped with a 10 meter rotor (swept area double that of the UM 70 machine). In other respects, the UM 100(Vms40) windgenerator is identical to the UM 70, which served as a basis for the development of the new machine.

As can be seen in the power and energy yield characteristics of the UM 100(Vms40) machine which have been included in the Appendix, the energy yield of this windgenerator should be nearly double that of a UM 70 installed at the same site in the GLENAN archipelago (especially during the summer months).

The price of the UM 100(Vms40) with its tower is about 22 % greater than that of the UM 70 with a tower of equal height. As maintenance costs for the two machines will be nearly the same, it follows that the energy supplied by the UM 100(Vms40) will be about 39 % less expensive than that produced by the UM 70.

The essential difference is that the lethal wind speed for the UM 70 windgenerator is about 110 m/s, while that of the UM 100(Vms40) is about 70 m/s. This parameter is sufficient to face the 30 year wind speed maximums encountered at numerous sites around the world; this period closely parallels the design life of the machine.

3.2. Towers

The towers used for these projects are of the same type used in Aerowatt installations during the last 20 years and are well adapted to sites with difficult access. The towers are guyed to permit erection and lowering of the windgenerator by means of a gin pole, thus avoiding the need for a crane and allowing all maintenance to be performed at ground level.

The tower height at PENFRET is 18 m. At SAINT-NICOLAS the tower height is 24 m. The towers for the windgenerators at BANANEC and DRENEC will both be 18 m.

3.2. Dual-conversion (AC/DC/AC) equipment

The first two projects, PENFRET and BANANEC, are equipped with what can be considered to be "classic" dual-conversion systems; i.e., the voltage of the battery bank being 120 V (60 lead-acid cells with positive tubular plates), the rectifiers include step-down transformers and the inverters include step-up transformers. The rated power of the associated inverter is 5 kVA in both cases.

While precise measurement of system efficiency has not yet been completed at PENFRET because of interruptions during the early phases of operation, information from an identical system (not financed by the Commission), which has operated without difficulty for 21 months, has shown that the efficiency of such a system is:

$$\tau \text{ (dual-conversion efficiency)} = E(aer)u/E(res) = 53\ \%,$$

where,

E(aer)u = AC energy at the rectifier inputs,
E(res) = AC energy delivered to the "grid" by the inverter.

It is important to recall that this dual-conversion efficiency includes:

- the efficiency of the rectifier with its step-down transformer,
- the efficiency of charge/discharge of the battery storage bank,
- the efficiency of the inverter with its step-up transformer.

Since it is not possible to modify the charge/discharge efficiency of the battery itself, Aerowatt International has sought to improve the efficiencies of rectification and inversion. To do this, Aerowatt International has specified development of inverters using an input voltage of about 300 VDC. This has led to a decrease in switching losses (the voltage drops in the semiconductors are independent of the operating voltage). Use of such a high voltage has also permitted elimination of the step-up transformer at the inverter output. The following results have been obtained:

- for pseudo-sinewave inverters of 10 and 25 kVA the no load power consumption has fallen to 15 W,
- for sinewave inverters of the same power ratings the no load power consumption is only 150 W.

This new generation of inverters has also permitted elimination of the step-down transformer at the the rectifier input. The 220 V rms windgenerator output is directly rectified to provide a charging current for the battery bank.

At present, results from installations using this type of inverter are not yet available from systems in the GLENAN archipelago. Results from an autonomous wind energy using this type of inverter at a joint, military/meteorological service base on EUROPA island (MOZAMBIQUE channel) show that over the last six months of operation the overall efficiency, τ, of this system has reached 72 %.

The systems at SAINT-NICOLAS and DRENEC will each be equipped with 10 kVA inverters of this new generation.

4. COMMENTS ON THE EXECUTION OF THE PROJECTS

The first two projects (PENFRET and BANANEC) were considerably delayed vis-a-vis the initial schedule presented to the Commission.

These delays were due to two factors: 1) the succession of Aerowatt by Aerowatt International required a certain amount of time before the new company was fully operational, 2) the client, CNG, experienced some difficulties in assembling the additional financing required to go forward with the projects.

The SAINT-NICOLAS project is advancing according to schedule and the first operating results should be available in the next few months

Given the experience gained in the first projects, the installation of the DRENEC system should proceed without difficulty.

At the PENFRET site, where the wind energy system has been operational since late 1986, the problems encountered to date have been essentially practical in nature. Since the base is not occupied on a permanent basis, delays have occurred in reporting operating problems. Most purely technical problems were avoided due to the parallel development and installation of similar systems for other clients.

Refinements in the equipment used for the GLENAN projects have primarily concerned adaptation of the electronic equipment (rectifiers) to accomodate higher than predicted voltage variations in the windgenerator output. Experience has shown that the voltage extremes observed (no load on the windgenerator and the machine operating in regulation mode) required a modification of the input circuits of the electronic equipment to permit such high voltages, especially for systems which are not actively used for long periods of time and where the battery bank can reach and remain at its maximum voltage during such periods.

A recent, routine maintenance visit to the PENFRET site revealed that, 18 months after commissioning, no extraordinary intervention was required.

5. SUMMARY OF TECHNOLOGICAL EVOLUTION DURING THE PERIOD 1983 - 1988

It is possible to compare the current cost of the system proposed in 1983 with that of a system using present technology. The energy available at the inverter outputs of each system is calculated assuming that both systems are installed on sites with average, annual wind speeds of 6.5 m/s and a wind speed distribution conforming to a Weibull's Law shape factor of k = 2.

This comparison will permit evaluation of the evolution in the unit cost of the energy produced by each system.

Description of 1983 equipment

UM 70/5000/BAT(50)NC windgenerator with SH 70-180 tower, AW/BAT 86/5000/120 battery charging and regulation unit, 60 cell (TSE 500) battery bank and 120/5000/221 SR inverter.

Theoretical energy production of the windgenerator	: 21,500 kWh
Dual-conversion efficiency	: 55 %
Energy theoretically available to the user	: 11,825 kWh
Ex-works cost of the equipment	: 267,700 FF (battery itself: 53,100 FF)

Description of 1988 equipment

UM 100(Vms40)/10000/BAT(50)NC windgenerator with SH 100(Vms40)-180 tower, AW/CEA/1/10 (308/10.000/223 SN 10 kVA inverter) dual-conversion unit, 155 cell (TSE 300) battery bank.

Theoretical energy production of the windgenerator	: 43,000 kWh
Dual-conversion efficiency	: 72 %
Energy theoretically available to the user	: 30,960 kWh
Ex-works cost of the equipment	: 371,790 FF (battery itself: 95,250 FF)

In both cases the annual maintenance expense is estimated to be 5000 FF/year. This comparison yields the following unit costs for the energy produced over the 20 year design life of the installations (asuming one replacement of the battery storage systems during this period):

- 1983 equipment : 1.76 FF/kWh
- 1988 equipment : 0.92 FF/kWh

6. COMPARISON WITH THE COST OF TRADITIONAL METHODS

The most recent information available to Aerowatt International concerning the cost of supplying energy using small diesel-based power plants comes from the Department of Electric Power Supply in the COOK ISLANDS. The operating authorities estimate that the real cost (including depreciation) of supplying energy on the outlying islands in this archipelago is currently (avril 1988) about NZ$2.00/kWh, or 7.5 FF/kWh.

Repeating the calculations of the preceeding paragraph, assuming a coefficient of 2.0 for the ex-works price of the wind energy systems, an amortization rate of 7.5 % and an annual maintenance expense of 10,000 FF (to take into account the geographic isolation), the full cost of supplying energy with autonomous wind energy power plants using 1988 is about 2.90 FF/kWh.

7. POTENTIAL MARKET FOR AUTONOMOUS WIND ENERGY POWER PLANTS

The comparison in the preceeding chapter of the energy production costs attainable with autonomous wind energy power plants and those observed in certain diesel-based systems must be expressed for systems of equivalent power ratings and energy yields.

The diesel grids on the outer islands of the COOK archipelago have, on average an installed generation capacity of about 35 kW and operate 12 hours/day. Assuming an average load of about 25 kW during the operating hours, the annual energy delivered to the grid is approximately 109,500 kWh.

As was stated above, the annual energy output of an autonomous wind energy power plant equipped with a single UM 100(Vms40) windgenerator (in the defined wind regime) will be about 31,000 kWh.

Aerowatt International has developed control equipment permitting as many as four windgenerators to be connected to the same battery bank. Further, high-efficiency inverters rated at 50 kVA are now available. An autonomous wind energy power plant equipped with three UM 100(Vms40)/10000/BAT(50)NC windgenerators, an AW/CEA/3/50 dual-conversion unit and a battery bank made up of 155 cells of 750 Ah capacity each has an ex-works price of 907,465 FF.

Applying the same augmentation coefficient used in the previous paragraph to account for shipping, customs duties and installation and assuming an annual maintenance expense of 30,000 FF, the unit cost of energy produced by such a system (about 93,000 kWh/year) will be approximately 3.11 FF/kWh.

These realistic evaluations of the energy production costs achieveable with autonomous wind energy power plants show that such costs can be significantly less than those observed in certain, traditional power systems. Real scale verification of these evaluations should permit the opening of significant markets at numerous isolated sites around the world.

NOTE : Criteria in force during past calls for tender issued by the Commission of the European Communities, rendered projects for autonomous wind energy power plants calling for the installation of more than one windgeneratôr ineligible for EEC financial support, whereas the modularity and enhanced system reliability resulting from the use of several identical machines are fundamental to this type of project.

Hélice............: UM70 PALES NACA 23021
Profil............: NACA 23021
Régulation........: UM70 NACA 23021 Ressort(5kW) Rapport 1/8
Multiplicateur....: GT5 RAPPORT 1/8
Génératrice.......: LS 132 - s

COURBE P(V) :

V (m/s)	P (kW)	V (m/s)	P (kW)	V (m/s)	P (kW)
0.5	+0.000	10.5	+5.898	20.5	+5.590
1.0	+0.000	11.0	+5.961	21.0	+5.566
1.5	+0.000	11.5	+5.961	21.5	+5.542
2.0	+0.000	12.0	+5.961	22.0	+5.520
2.5	+0.000	12.5	+5.949	22.5	+5.499
3.0	+0.000	13.0	+5.929	23.0	+5.479
3.5	+0.000	13.5	+5.910	23.5	+5.459
4.0	+0.000	14.0	+5.891	24.0	+5.439
4.5	+0.262	14.5	+5.872	24.5	+0.000
5.0	+0.559	15.0	+5.853	25.0	+0.000
5.5	+0.931	15.5	+5.830	25.5	+0.000
6.0	+1.355	16.0	+5.807	26.0	+0.000
6.5	+1.826	16.5	+5.783	26.5	+0.000
7.0	+2.346	17.0	+5.760	27.0	+0.000
7.5	+2.909	17.5	+5.736	27.5	+0.000
8.0	+3.490	18.0	+5.712	28.0	+0.000
8.5	+4.081	18.5	+5.687	28.5	+0.000
9.0	+4.670	19.0	+5.663	29.0	+0.000
9.5	+5.245	19.5	+5.638	29.5	+0.000
10.0	+5.806	20.0	+5.614	30.0	+0.000

PRODUCTIBILITE POUR UNE DISTRIBUTION DE WEIBULL

	k=2		k=3		k=4	
Γ(1+1/k)	0.886227		0.892979		0.906403	
V (m/s)	Pm (kW)	Ea (kWh)	Pm (kW)	Ea (kWh)	Pm (kW)	Ea (kWh)
2.0	0.015	131	0.001	5	0.000	0
2.5	0.072	631	0.013	112	0.002	16
3.0	0.194	1698	0.069	602	0.027	234
3.5	0.385	3375	0.194	1698	0.114	996
4.0	0.638	5588	0.395	3462	0.282	2468
4.5	0.935	8194	0.668	5855	0.530	4640
5.0	1.259	11025	1.003	8784	0.849	7435
5.5	1.590	13926	1.383	12113	1.228	10761
6.0	1.916	16781	1.788	15666	1.655	14500
6.5	2.227	19506	2.199	19261	2.111	18496
7.0	2.517	22052	2.596	22741	2.574	22552
7.5	2.784	24391	2.968	25997	3.022	26469
8.0	3.026	26510	3.306	28963	3.435	30089
8.5	3.242	28402	3.609	31614	3.804	33321
9.0	3.432	30066	3.875	33948	4.124	36130
9.5	3.596	31505	4.108	35983	4.397	38521
10.0	3.735	32722	4.308	37741	4.627	40529
10.5	3.850	33724	4.481	39250	4.817	42195
11.0	3.941	34521	4.627	40531	4.973	43567
11.5	4.009	35123	4.749	41602	5.101	44688
12.0	4.058	35544	4.848	42470	5.205	45599
12.5	4.087	35799	4.925	43141	5.289	46330

AW 86.04.19 Mod. 86.09.06

Aérogénérateur UM 70/ 5000/BAT-NC

COURBES P(V) et Pm(Vm) (Weibull, k=2)

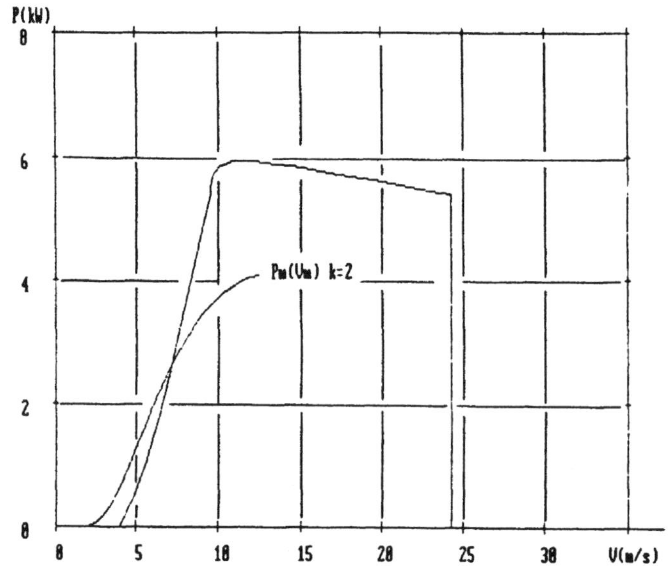

Nota : Les puissances et les énergies sont données en courant continu. Elles peuvent différer légèrement selon la tension batterie.

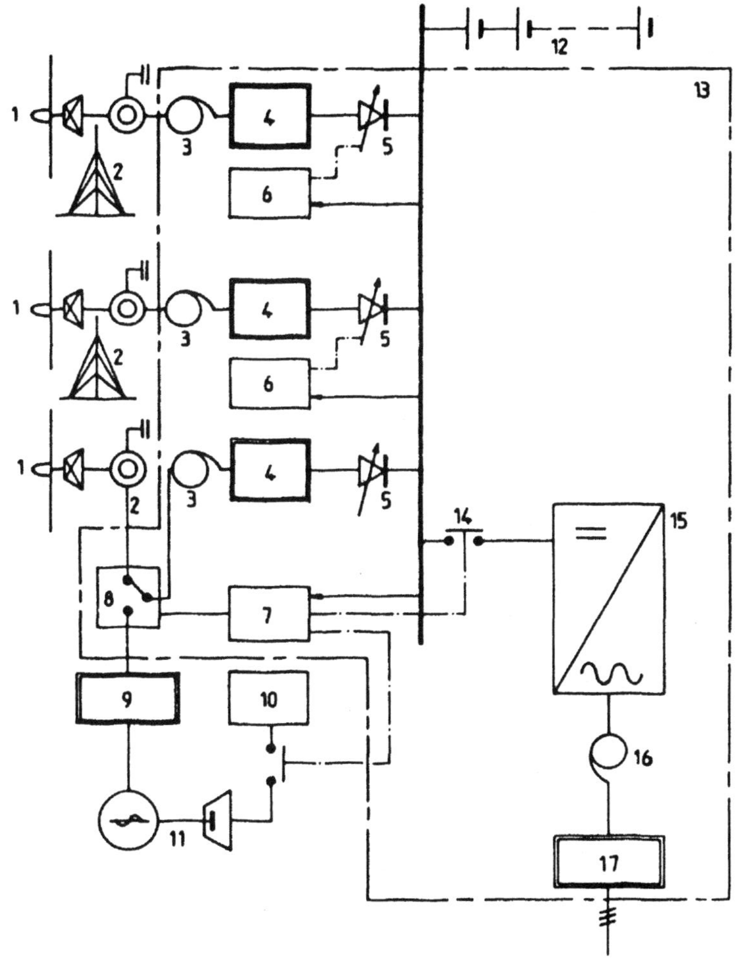

AUTONOMOUS WIND ENERGY POWER

-=-=-=-=-=-=-

SINEWAVE INVERTER

-=-=-=-=-=-=-

BLOCK DIAGRAM

N°	QUANTITY	DESCRIPTION
1	1 - 4	WINDGENERATOR
2	1 - 4	WINDGENERATOR TOWER
3	1 - 4	(OPTION A) STEP-DOWN AUTOTRANSFORMER (380 V/220 V)
4	1 - 4	WATT-HOUR METER
5	1 - 4	CONTROLLED RECTIFIER
6	0 - 3	AWI 1380 CIRCUIT CARD WITH MES OPTION
7	1	AWI 1380 CIRCUIT CARD WITH GES AND MES OPTIONS
8	1	COMMUTATION CONTACTOR : WINDGENERATOR/STAND-BY GENERATOR SET
9	1	WATT-HOUR METER
10	1	STAND-BY GENERATOR SET START-UP CIRCUIT
11	1	STAND-BY GENERATOR SET
12	1	BATTERY BANK - 155 CELLS (2 V EACH)
13	1	AW/CEA/1-4/10, 25 or 50 DUAL CONVERSION ENERGY PROCESSING UNIT
14	1	DISCHARGE LIMITATION CONTACTOR
15	1	INVERTER : AW/310/10000/223 SN OR AW/310/25000/223 SN OR AW/310/50000/223 SN
16	1	(OPTION B) STEP UP AUTOTRANSFORMER (220 V/380 V)
17	1	WATT-HOUR METER

Project Nr 306/84 DE

DEVELOPMENT AND CONSTRUCTION OF A MODULAR SYSTEM FOR AN AUTONOMOUS ELECTRICAL POWER SUPPLY ON THE IRISH ISLAND OF CAPE CLEAR

Dipl.-Ing. R. Grebe
Dipl.-Ing. G. Cramer
SMA Regelsysteme GmbH
Hannoversche Str. 3, 3501 Niestetal 1
Federal Republic of Germany

1. Introduction

During the last years, a combination of wind energy converters (WECs) and diesel engines has proved to be a very interesting application for the supply of electrical power in remote regions.

Within the scope of this demonstration project a modular wind/diesel/battery combination was developed which allows to switch off the diesel engine in time of good wind conditions. A short time-battery storage realizes minimum diesel running times and a maximum fuel saving. Moreover, the use of a battery allows to reduce the number of start-stop cycles considerably in case of low wind velocity.

This combination with its minimum of electrical installations, and high efficiency, makes possible an optimum supply of electrical consumers not connected to the public grid. The quality of electrical power supplied by this system is for all operation modes (parallel operation of WECs and diesel generating sets, parallel operation of storage unit and WECs, or supply by the diesel generating set alone) comparable with small or medium diesel power stations regarding constancy of voltage and frequency.

2. System design

The supply system consists of two wind energy converters of AEROMAN type with a nominal power of 30 kW each, a diesel generator unit made by MAN with a nominal power of 72 kW and a battery storage array with a capacity of 100 kWh made by Hagen Batterie AG. In former times the electric energy was generated by two diesel-generator sets with nominal powers of 100 kW and 164 kW. The energy is distributed via a 10 kV overhead line.

The annual energy consumption is about 250.000 kWh. The day load curve shows a minimum of about 15 kW early in the morning and a maximum of about 80 kW to 120 kW in the evening. The system components have been selected according to the following requirements:

- guaranteeing a safe electric power supply
- high total efficiency
- high availability
- low maintenance expenses
- high life-time.

The principle system design is shown in the block diagram, figure 1.

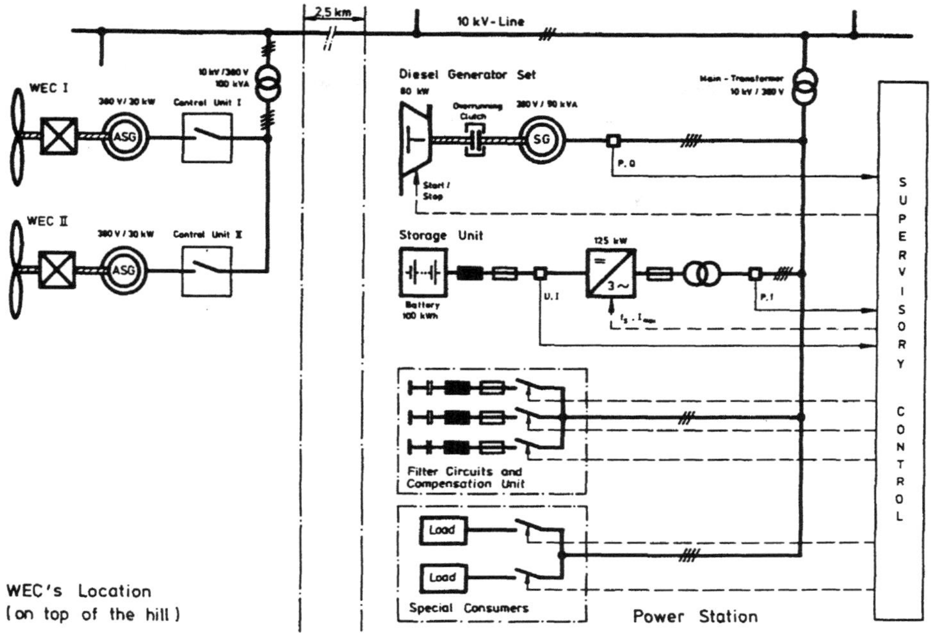

Figure 1: Design and dimensioning of the autonomous electrical power supply system for Cape Clear Island

2.1 Diesel-generator set

The diesel-generator set consists of a boosted MAN diesel engine with a nominal power of 80 kW which is coupled with the 90 kVA synchronous generator by means of an overrunning clutch.

In this way the synchronous generator works as a rotating phase-shifter when the diesel is off, and takes over voltage control and the supply of reactive power. During low-load periods and high wind power supply it is possible to shut the diesel down completely, and to maintain supply by means of the wind energy converters and the battery storage alone.

Figure 2: Diesel-generator set (90 kVA)

2.2 Wind Energy Converters

The WECs of AEROMAN type installed within the scope of this demonstration project dispose of a high-standard speed and power control. This allows an optimized parallel operation of the WECs and the small diesel-generator unit. Moreover, the WECs are able to take over frequency control even when operating solely without being supported by the diesel engine (but supported by the battery with its converter).

The WECs are standard types for grid-parallel operation, and are equipped with asynchronous generators. The nominal power output of each wind energy converter is 30 kW at a wind velocity of 11.4 m/sec.

Figure 3: View of the wind energy converters AEROMAN on Cape Clear

2.3 Battery Storage System

The system is equipped with a battery storage to avoid unnecessary frequent startings of the diesel engine caused by fluctuations of wind velocity or consumer demands.

The installed OCSM-battery made by "Hagen Batterie AG" consists of 165 single cells, with a total nominal capacity of 300 Ah. This corresponds to an energy of about 100 kWh.

A line commutated converter with a rated power of 125 kW takes over the battery charging and the supply of the isolated grid. This stands out for simple construction and low price.

Furthermore, the installation of a storage unit makes possible the operation of the diesel engine in more suitable power ranges, because e.g. load peaks usually requiring the starting of the diesel can be checked now by the storage unit. Hence, the diesel engine may be sized quite small, because it has not to cope with the load peaks.

Figure 4: Battery storage (100 kWh)

3. Supervision Unit

The microprocessor-equipped operation control system choses the optimum operation mode as regards economy and the supply safety, depending on the actual power output of the plant, the battery's charging state, the actual load conditions, and the expected consumer demands. The tasks of the supervisory control system are as follows:

- Diesel Start / Stop Control
- Battery Charging Control
- Battery Charging Situation Monitoring
- Frequency Control
- Compensation Control
- Failure Check and Display

The control of the power supply ratio between different supply units of the system works on the basis of a frequency/power characteristic, this means that the actual grid frequency is taken as base value for the plant's supervision unit.

All necessary switching devices, protection and monitoring devices, and the hardware of the supervisory control computer are installed in the switch cabinet, see figure 5.

Figure 5: Switch cabinet with converter and supervision unit

4. Operation Modes

The supply system on principle disposes of 4 different operation modes. The transition from one operation mode to another occurs without any interrupt, so that the consumers remain uninfluenced. Which operational mode is realized, always depends on the wind conditions, the demand, and the battery charge state.

Parallel operation of WECs, diesel engine, and battery storage
During periods with low wind velocity and high consumer demands.

Parallel operation of diesel engine and battery storage
During periods with unsuitable wind conditions.

Parallel operation of WECs and battery storage
During periods with sufficient wind velocity.

Single operation of the battery storage
During short periods with low wind velocity and low consumer demands.

The synchronous generator acts as phase shifter during operation modes without diesel engine. Furthermore, it takes over voltage control and the supply of reactive power. The battery converter during all operation modes takes over frequency control and the active power's distribution.

5. Measurement Program

A one-year measurement program was carried out to estimate the availability and economy of the installed supply system. Moreover, the results should allow further optimization of particular plant components and of the supervision strategy.

All relevant measured values were read with a sampling rate of 10 Hz per channel by means of a computer aided measurement system. During nominal operation these measured values were filed on floppy disc in form of 10-minute mean values. The minimum and maximum values of each 10-minute interval were filed, too.

Daily records were made which show the prepocessed measured values graphically and in form of tables. Figure 7 shows the daily record of 30th August, 1987.

In case of a supply system failure (significant values exceed their programmable limit values), all measured values are filed with a high sampling rate for a minute and are displayed graphically.

The measured values filed on floppy disc were evaluated additionally by SMA. The hardware bases on microprocessor modules which are compatible to SMP-bus modules. The multitasking Concurrent DOS serves as operating system. The user exclusively communicates via menus.

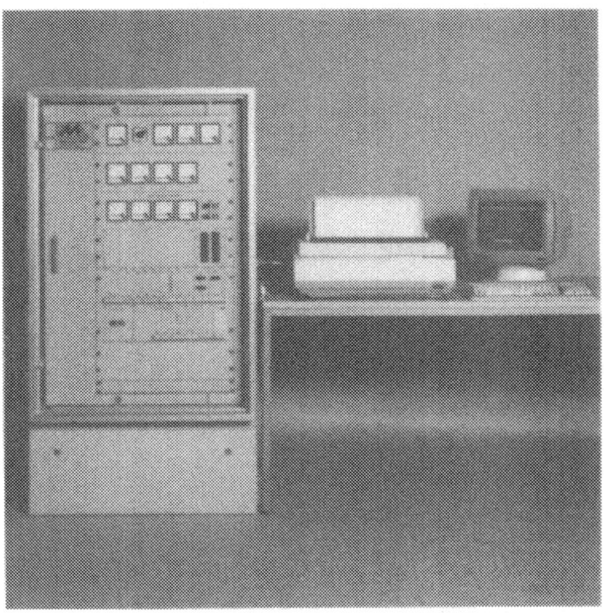

Figure 6: The Measurement System

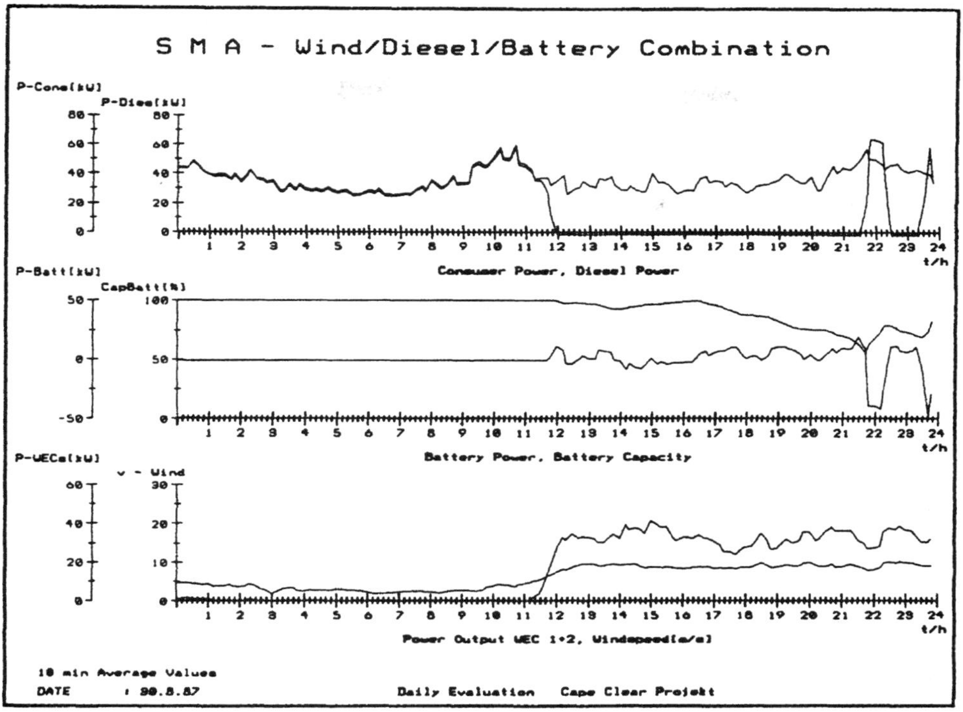

Figure 7: Graphical representation of the power output of one day

6. Measurement Result

At the beginning of the project only both wind energy converters were installed and did operate parallel to the existing diesel-generator sets in fuel saver operation. The WECs did feed electrical power into the island grid, dependent on the wind conditions. Thereby, the output power of the WECs had to be reduced in case of significant unloading of the diesel engine. About 1/3 of the consumer demand was supplied by the WECs in this operation mode.

In the second phase, starting at the end of August 1987 the advanced operation mode could be realized, by installing the disconnectible diesel-generator set and the battery storage. During the first 3 months of operation the WECs supplied about 51.7 MWh electrical energy, which is about 70% of the total consumer demands.

This means an annual fuel saving of about 60 tons. The diesel running time could be reduced to 35%, because in case of proper wind conditions the diesel engine is switched off. Hence, the life time of the diesel prolonges correspondingly.

The last figure shows the proportion of wind energy to the energy produced by the diesel-generator in November 1986 and November 1987. The comparision shows that the share of wind energy increased from 36% in the fuel saver operation to 72% in the operation with the complete system. This illustrates the advantages of the advanced system configuration.

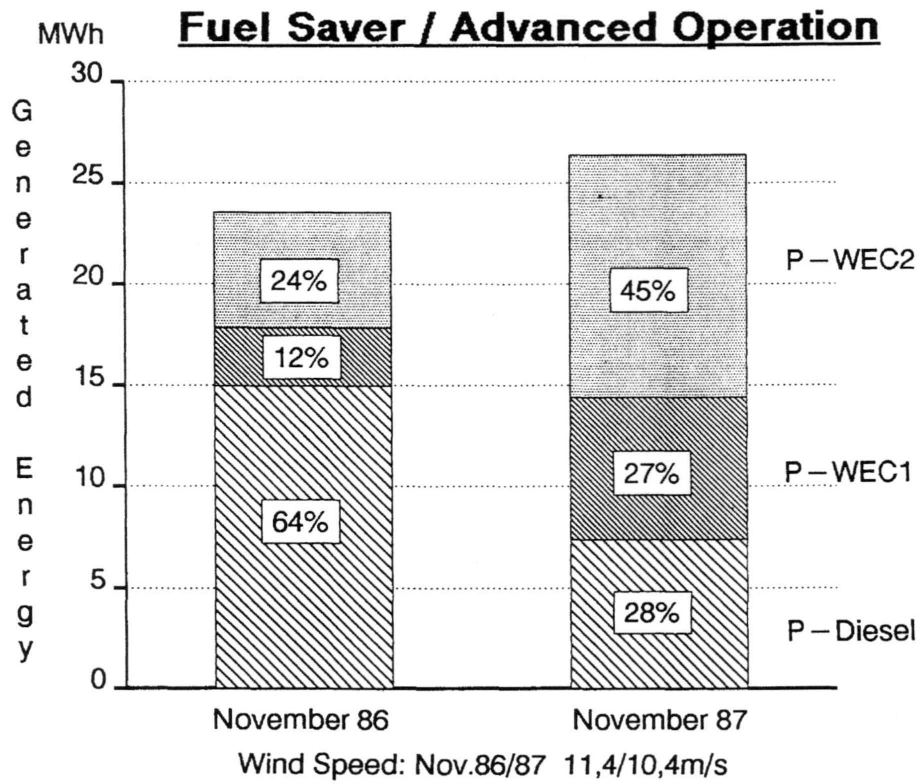

Figure 8: Proportion of the wind energy to the energy produced by the diesel-generator in November 86 and November 87

Project Nr 324/84 DE

WINDENERGY CONVERTER FOR AEROBIC TREATMENT OF SEWAGE

G. Huppmann
Messerschmitt-Bölkow-Blohm GmbH
Energy and Process Technology
P.O.B. 80 11 09
8000 München 80

Summary

An existent biological sewage treatment plant is supplied with electrical energy produced by a combined wind/diesel power supply unit consisting of an MBB MONOPTEROS 15, a 25 kVA diesel generator and the necessary master control unit.

The main objective of this demonstration project is the design, construction, installation and test of the autonomous wind/diesel system and the evaluation of economical data for a wind/diesel driven sewage treatment plant in a remote area avoiding high cabling costs and saving conventional energy. The plant is under operation since December, 1987 and gained further developments, such as a winddriven seawater desalination unit with a highly increased fraction of energy produced by wind.

Main Objectives

The main objectives of this demonstration project wind energy converter for aerobic treatment of sewage were

- to feed an existent biological sewage treatment plant with electrical energy from wind, using the MBB wind energy converter MONOPTEROS 15
- to test the energy coproduction of the combined wind/diesel power supply unit MONOPTEROS 15 wind/diesel
- to evaluate the economical data of an autonomous electrical power supply for water and sewage treatment units, which are very often located far away from electrical grids.
  High cabling costs can be avoided and conventional energy can be saved.

The chosen sewage treatment plant is located in the MBB-airfield Lemwerder and discharges the MBB plants of Lemwerder near Bremen.

The aerobic sewage treatment is installed in two bassins and works with two air compressors. The amount of sewage to be treated is about 40.000 m^3/year.

Both compressors need 11 kW, 5.5 kW each.

Due to the necessity of continuous working of one compressor at least the electrical energy amount reaches 132 kWh per day in minimum. The plant is connected to the grid by a 1000 m cable.

Instead of this electrical line from the grid an autonomous electrical supply consisting of a wind energy converter MONOPTEROS 15 and a diesel generator has been designed, erected and began to be tested within this project. A 100% electrical supply is the target and it seems to be possible, that it will be realized at the end of the demonstration project.

Structure of the project

The project was executed in four successive phases:

1. System definition and design.
   Preparation of infrastructure of the sewage treatment plant.

2. Preparation of the MONOPTEROS 15 for autonomous operation in combination with a diesel generator without any grid connection. The MONOPTFROS 15 was available for grid connection only before.

3. Procurement, erection, integration of all components of the combined power supply unit. Connection to the treatment plant. Commissioning.

4. Continuous test operation for 1 year. Evaluation of the number of operational hours, produced electrical energy both of WEC and diesel. Control and switch behaviour, frequency changes etc.

System Description

The system consists of 3 main components:

- wind energy converter MONOPTEROS 15 with internal control unit

- diesel generator with internal speed control

- master control unit

The MONOPTEROS 15 with a rated performance of 22 kW operates in parallel to the diesel generator (20 kW).

The mechanical coupling between the synchronous generator of the diesel engine and the diesel itself is made by a overrunning clutch in such a way, that the generator is allowed to rotate with a higher speed than the diesel but not vice versa.

Due to this clutch, the diesel can be stopped if enough power coming from wind is available to fulfill the requirements of the actual loads.
In this case the synchronous generator works as a rotating phase lifter to produce the necessary reactive power.

If the diesel is stopped, the frequency of the whole system will be controlled by fast switch-on and -off of resistors by thyristors.
Three operational modes are possible:

- operation of diesel alone
- operation of MONOPTEROS 15 parallel to diesel
- operation of MONOPTEROS 15 alone.

Fig. 1 shows the flux diagram of the system, fig. 2 describes the Master Control Unit.

Technical Data

The technical date of the main components of the combined wind/diesel power supply unit are as follows:

Wind energy converter:

Type:	MONOPTEROS 15
Regulation:	Stall controlled
Performance:	22 kW
Rotordiameter:	12.5m
Rated speed:	100 rpm
Hubheight:	15 m
Number of blades:	1
Material:	Fiberglass-reinforced plastic
Switch-on speed:	4.5 m/sec
Switch-off speed:	20 m/sec
Survival speed:	60 m/sec in parking mode
Total weight:	1540 kg

Dieselgenerator:

Type:	D-4010. Standard Aggregatebau GmbH, Hamburg
Performance:	25 kVA / 20 kW
Generator:	3 phase synchronous generator, collector-free, Type A.v.Kaick
Voltage:	380 / 220 Volt, 3 phase, 50 Hz, cos phi .8
Speed:	1500 rpm
Clutch:	Overspeed clutch, Ringspann, Type FKHL 24 ATR
Diesel:	Deutz, Type F3L912, air-cooled, 34 PS, DIN 6271

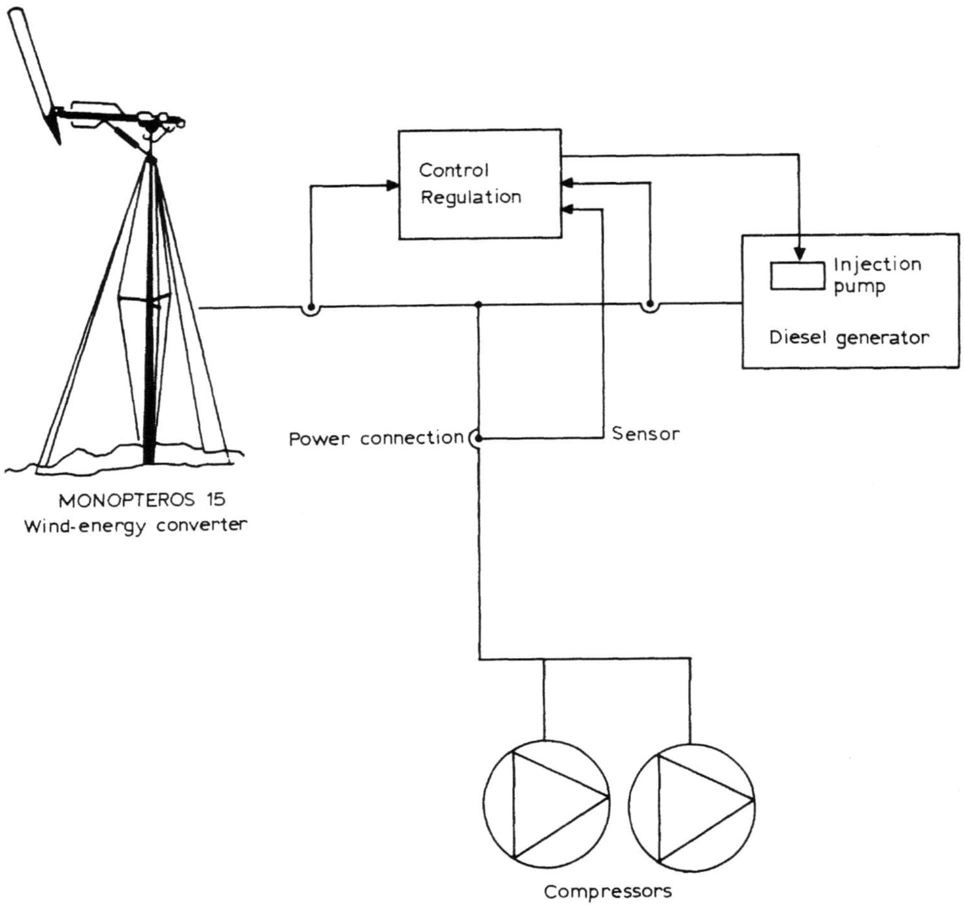

Fig. 1: Flux diagram of the Lemwerder Plant

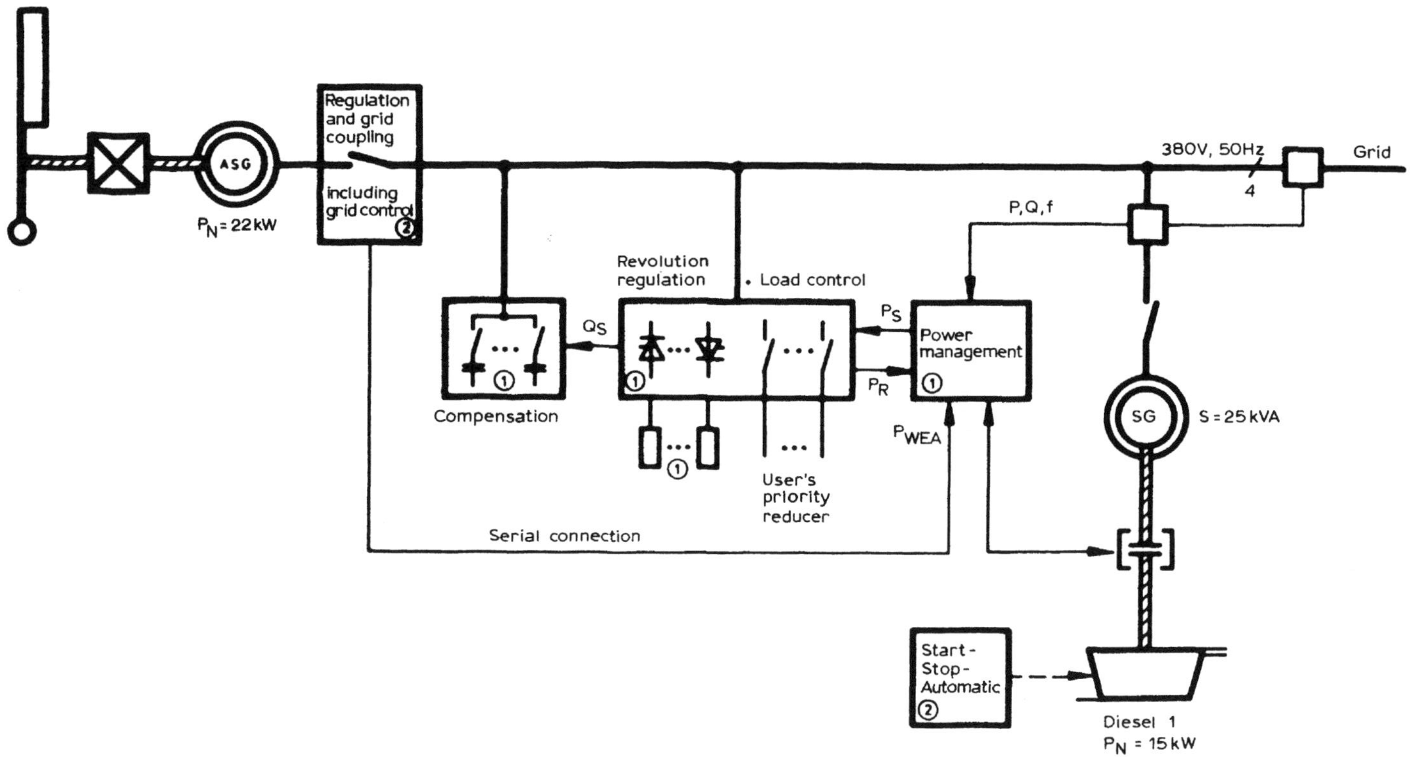

Fig. 2: Master Control Unit

Fig. 3 shows the power curve of the MONOPTEROS 15, fig. 4 the annual energy yield and fig. 5 the performance distribution of the plant at Lemwerder. Fig. 6 the power curves of the advantaged MONOPTEROS 15 continuous pitch controlled, fig. 7 the annual energy yield and fig. 8 the related performance dstribution of a winddriven seawater desalination plant with this MONOPTEROS 15 Pitch at a 6.5 m/s location, as it is at Marsa Matrouh / Egypt.

Status

The combined wind/diesel power supply unit MONOPTEROS 15 W/D is in operation since 3.12.1987.

Till the time being we got the following problem:

- Fault in the internal diesel control system. The diesel did not stop in case of enough wind and enough power production of the WEC.

The evaluation of the operational data has been started.

Status at 22.4.1988:

Operational hours of the plant:
3024 h (100%)
Operational hours of the WEC:
1050 h (35%)
Operational hours of the diesel:
1821 h (60%)
Energy produced by combined power supply:
12195 kWh
by WEC
2043 kWh = 17%
by Diesel
10152 kWh = 83%.

The rather poor wind fraction of 17% is mostly caused by the following facts:

- The fault of the diesel control system no stopping the diesel in case of enough wind

- the wind regime at Lemwerder

- problem with finding the optional adjustment of the control parameters. Especially the minimum diesel running time in combination with minimum power output of WEC. This couple of parameters has lead to short diesel stop periods, caused by frequent drop in of windspeed under the threshold of diesel start operation.

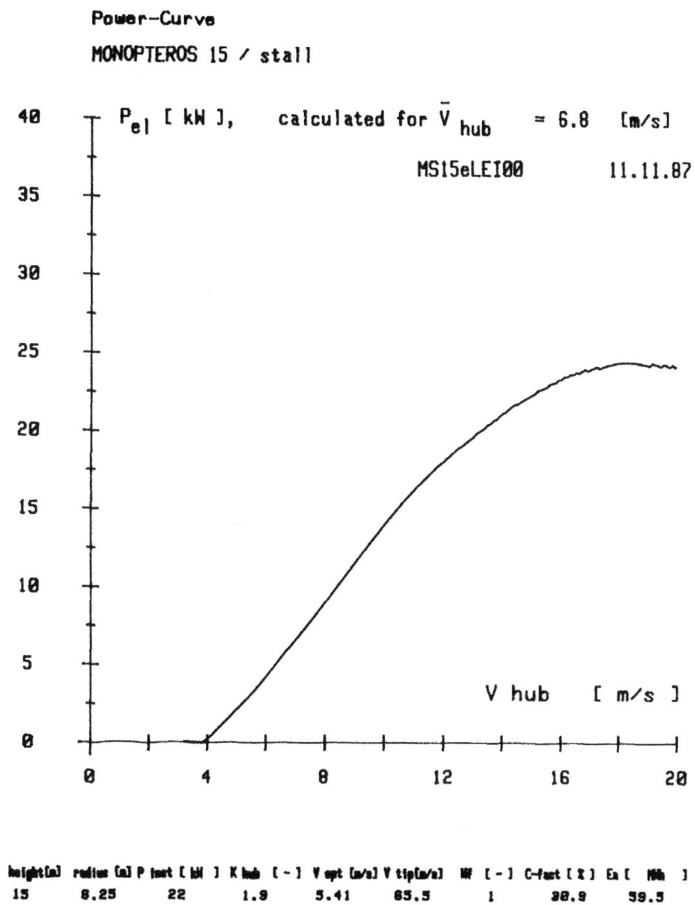

Fig. 3 : Power Curve MONOPTEROS 15 Stall

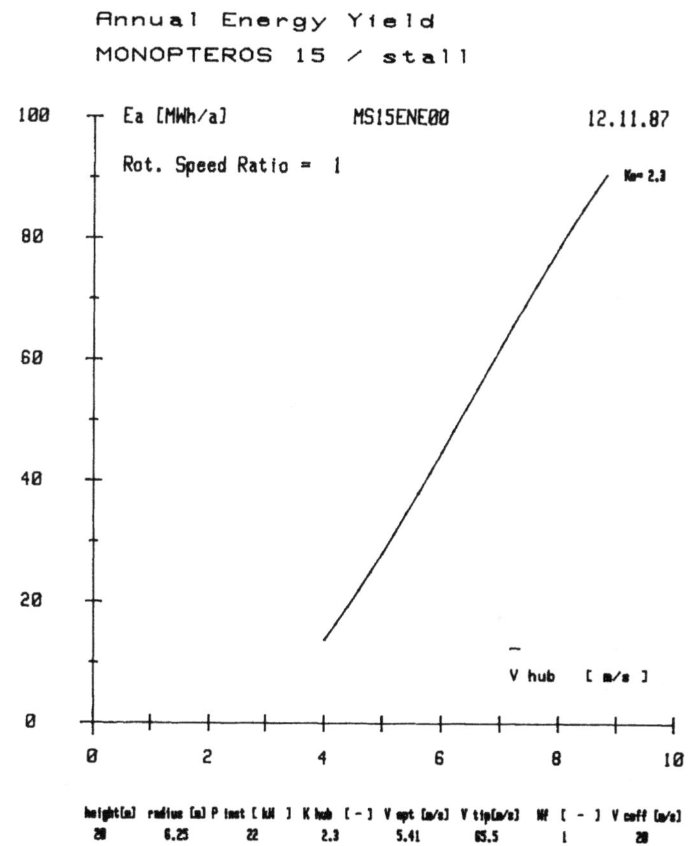

Fig. 4 : Annual Energy Yield of MONOPTEROS 15 Stall

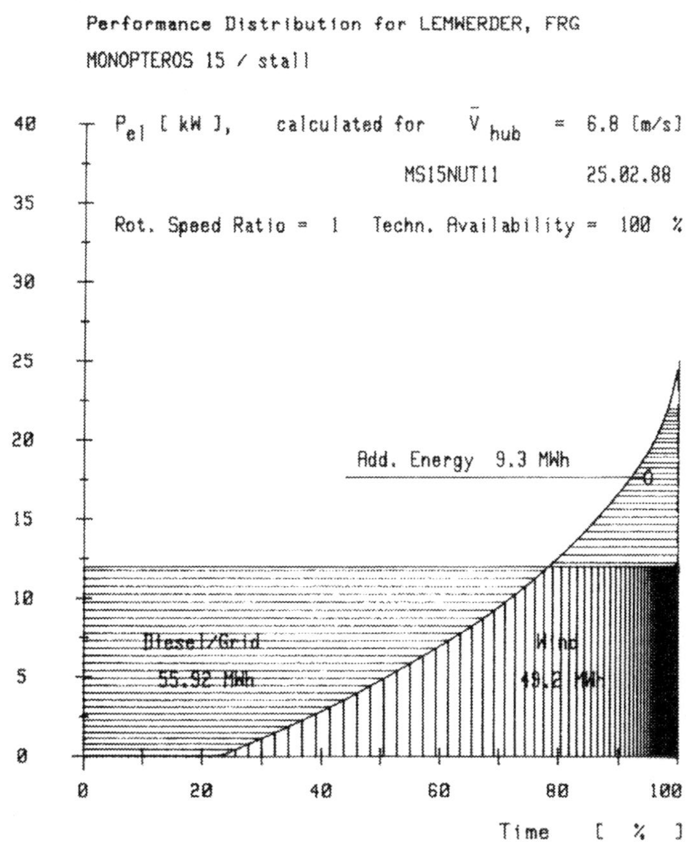

height[m]	radius [m]	P inst [kW]	K hub [-]	V opt [m/s]	V tip[m/s]	Wf [-]	C-fact [%]	Ea [MWh/a]
15	6.25	22	2.1	5.41	65.5	1	40.8	49.2
							30.4	58.5

Fig. 5 : Performance Distribution Lemwerder

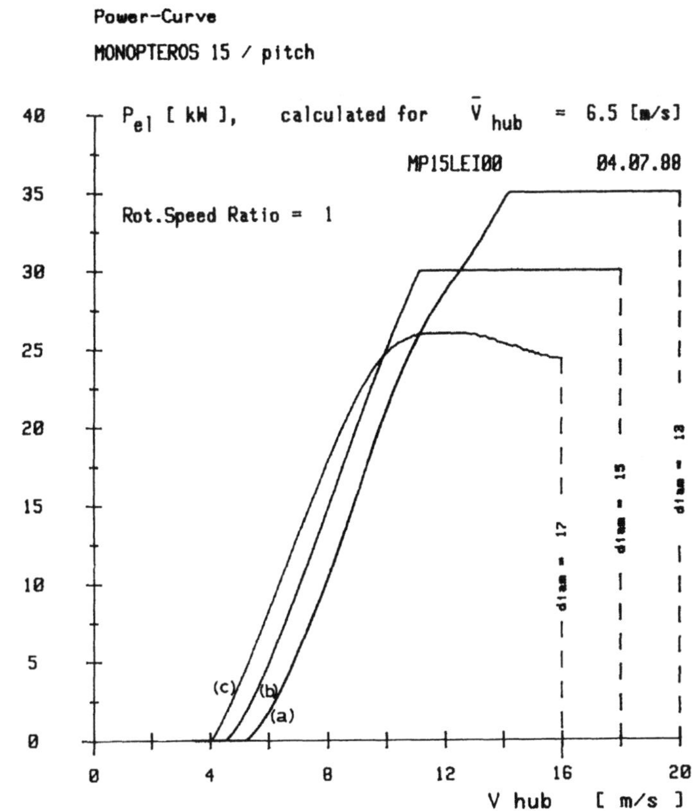

	height[m]	radius [m]	P inst [kW]	K hub [-]	V opt [m/s]	V tip[m/s]	Wf [-]	C-fact [%]	Ea [MWh]
(a)	20	6.5	35	2.03	7.10	79	1	22.1	68
(b)	20	7.5	30	2.03	6.17	79	1	32	84
(c)	20	8.5	26	2.03	5.1	79	1	40	91

Fig. 6 : Power Curves MONOPTEROS 15 PITCH

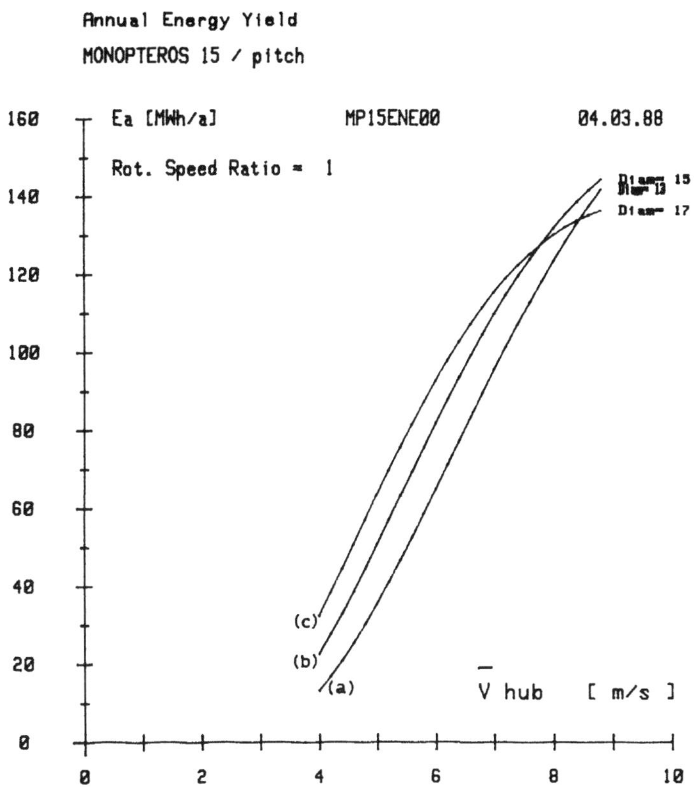

	height[m]	radius [m]	P inst [kW]	K hub [-]	V opt [m/s]	V tip[m/s]	Nf [-]	V coff [m/s]
(a)	20	6.5	35	2.1	9.09	100	1	20
(b)	20	7.5	30	2.1	7.81	100	1	18
(c)	20	8.5	26	2.1	6.45	100	1	16

Fig. 7 : Annual Energy Yield MONOPTEROS 15 PITCH

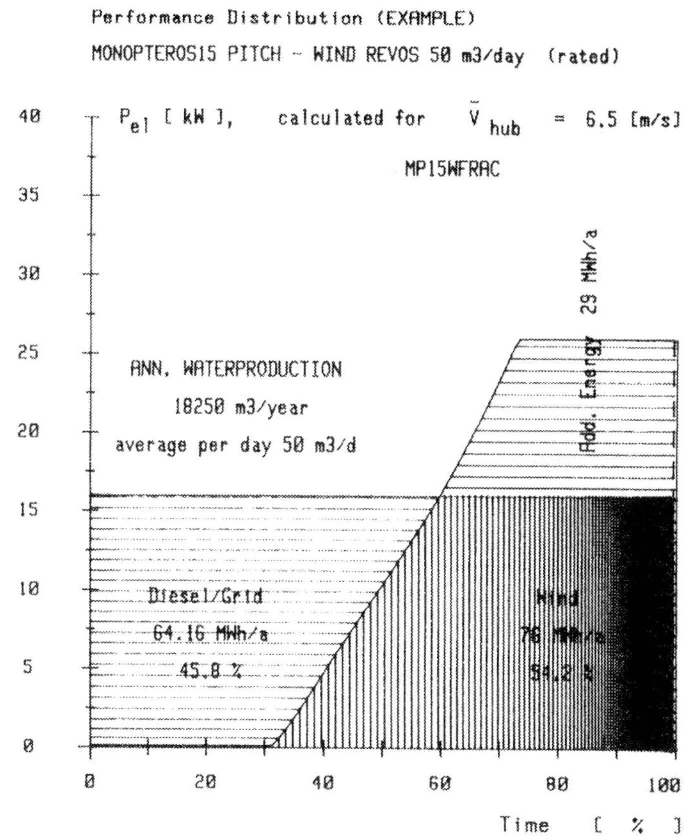

height[m]	radius [m]	P inst [kW]	K hub [-]	V opt [m/s]	V tip[m/s]	Nf [-]	C-fact [%]	Ea [MWh]
20	8.5	26	2.1	6.45	100	1	54.2	78
							46.1	105

Fig. 8 : Performance Distribution MONOPTEROS 15 PITCH

Further developments

The described project gained further developments during its time of realization.

- WINDDRIVEN WATERPUMPING STATION

    This is a demonstration project at Marsa Matrouh, Egypt. It consists of a stand-alone version of the MONOPTEROS 15 STALL and a cascade of 3 waterpumps.

- WINDDRIVEN SEAWATER DESALINATION PLANT "WIND REVOS"

    A commercial plant in Marsa Matrouh, Egypt, also. The power plant works following the same principle as the described plant at Lemwerder and produces the electrical energy for a reverse osmosis seawater desalination unit. The plant consists of a MONOPTEROS 15, a 25 kW diesel generator and the reverse osmosis unit REVOS, designed and delivered by Kraftanlagen Heidelberg.

    The output of the plant is 25 m^3 per day potable water according to the standards of WHO.

    The plant is in operation since end of 1987 and has been accepted by the customer in February, 1988.

    The wind energy fraction reached till now is approx. 52.5%.

Project Nr 370/83 HE

MYKONOS ISLAND WIND PROJECT

J.L.TSIPOURIDIS, A. ANDROUTSOS, G.VERGOS
A. KORONIDIS, P. PLIGOROPOULOS

P.P.C./DEME, NAVARINOU 10, ATHENS 106 80, GREECE

Summary

A 108 KW wind turbine has been installed and is operating on the island of Mykonos since December 1986 connected to the island's diesel-grid. Civil works, manufacture, grid works, transportation, erection and commissioning were completed in about 9 months. As the installation site (Fanari) has an average wind speed of 11m/s, the WT is operating often close to its rated output. Operational data to date show availability of 64,3%,capacity factor of 35,1%,while total production is 436.263KWh (8.12.86-31.3.88).Regarding maintenance and parts-replacement there have been a few problems leading to production losses,which are expected to be eliminated through better coordination of related activities.

1. INTRODUCTION

The aim of this demonstration project was to prove the reliability of wind energy conversion by integrating a medium sized unit in a small diesel-grid, in an island with insufficient infrastructure but with excellent wind conditions. The final outcome, determined by parameters such as costly electricity production, high wind potential, lack of infrastructure,access difficulties and environmental restraints, will establish the viability of wind energy exploitation in general. The result in connection with the degree of penetration that can be optimally achieved will determine the extent of wind energy exploitation in the Aegean islands.

2. DESCRIPTION OF WT'S COMPONENTS

The main technical data of the Danish MICON 108 KW unit is as follows:

Tower : 24-edged tubular, 3-sections hot-dip galvanised 22m height.

Rotor : 23m hubheight - 3MAT blades, 19.3m diameter, $293m^2$ swept area, fixed pitch (twist angle 12 deg),upwind stall control, parachute-brakes.

Gearbox : Kumera OY (FL), two stage (mainshaft 42 rmp generators 1015 rpm approx.)

Generator: Elin , MKG - 9 , 3 phase asynchronous at 108KW, 3X380V and 50Hz. Protection IP 54, insulation F, 1015rpm.

Yaw: Two stage worm/planetary, driven by 0,55KW motor , ratio 960: 1

Safety: fail -safe electromagnetic disk-brake, self activating parachutes at wing speed above 51 rpm, overspeed control, automatic yawing of nacelle, Automatic stop at unbalance, brake-pads warning light, Thermal control of generator and gear box causes WT to stop, cable untwisting procedure automatic.

Performance: Cut in 4 m/s, cut-out 28 m/s, Nominal output at at 14 m/s, Wing tip speed 43 m/s.

Control: local and remote, all functions/errors supervised , microprocessor registers performance data. Electonic engaging equipment restricting current innpulses on egagement to max 1.1 X full load of the generator. Capacitors keep cos $\Phi = 0.90$. Over voltage protection against lightning. Automatic start-up after grid failures and manual remedy of other errors.

3. Project Management

The project was realised in a period of 9 months from the signing of the Consortium Agreement to Commissioning. This is considered as exceptionally quick if the problems associated with working conditions in the islands are taken into consideration . Furthermore the fact that the project was already delayed for more than a year due to the original partners withdrawal from the project dictated rapid completion of works.

Civil works were carried out by a local subcontractor while grid extension substation works and telephone line construction works were carried by the local branches of PPC and Greek Telecommunications Organisation respectively. A PPC crane was ferried over for erection. Major difficulties were encountered in transporting the WT to the site through the narrow, winding and crowded (due to the summer tourist season) Mykonos' roads, and in erecting the machine due to the prevailing high winds (above the specified for safe-erection speed of 8 m/s.)

There was a delay in the commissioning due to replacement-repair work that had to be carried out in the parachutes and the yawring bolts, and due to problems in the electrical connections to the grid, but still within the time schedule.

MYKONOS ISLAND

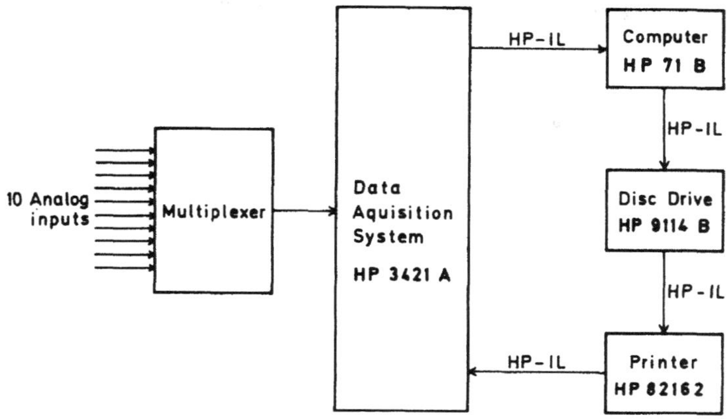

Figure 1. DATA ACQUISITION BLOCK DIAGRAM

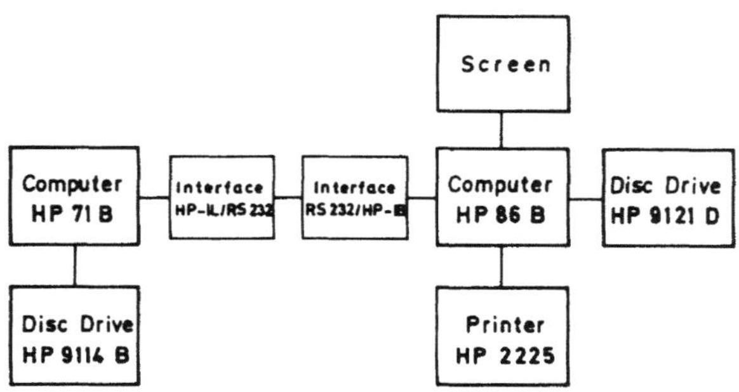

Figure 2. SYSTEM FOR ELABORATION DATA AND ANALYSIS

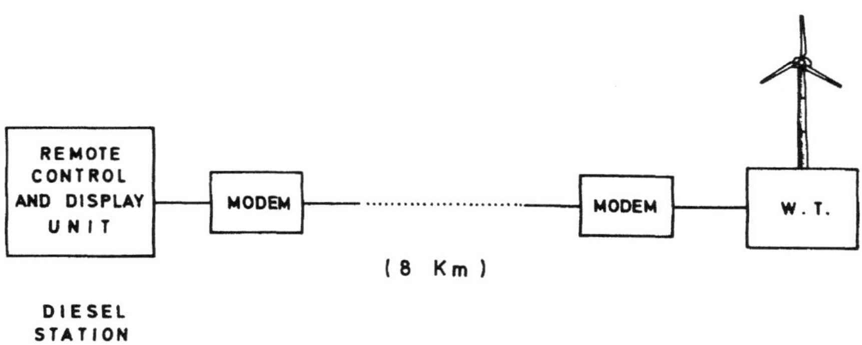

Figure 3. REMOTE CONTROL BLOCK DIAGRAM

Since the last reporting period further work was carried relating to the data aquisition and the remote control system.

PPC personnel designed the data monitoring system, working out the software and specifying the hardware required to monitor all the important operation parameters. Equipment fit for professional data monitoring applications were selected as access difficulties to the island rendered reliability of operation of the outmost importance. Further more the demand for reliable readings dictated high precision instruments.

The data acquisition system technical description is as follows : (Figures 1,2 and 3 and Table I).

This system has been designed with Hewlett-Packard equipment. The HP 3421A Data Aquisition unit consists of a main frame with an integrating multimeter and 2 slots for assemblies . Also it has an integrating 5 1/2 - 4 1/2 - 3 1/2 digit multimeter, frequency counter and thermocouple compensation. The multiplexer assemblies provide 10 channels for scanning imput signals.

The HP 71B Handheld Computer is connected to the Data Aquisition unit via a HP - IL Link which is a bit serial interface designed for low cost battery operable systems.

In HP - IL systems devices are connected by two-wire cables leading from the output port of one device to the imput port of the next, until all devices form a closed loop.

The HP 71B computer includes a quartz crystal clock and can turn on the HP3421A unit, trigger to scan a list of up to 30 inputs, instruct a mass memory to store the reading and then power down the entire system until the next time.

The HP 9114B portable Microfloppy is employed as mass storage media .

The data scanning of the required measurements is done on a periodic basis with a time interval of the order of a few seconds (i.e. 10 seconds). By an aprropriate subroutine running in the HP 71B computer the average value of the sampled measurements is then calculated and stored in the computers RAM every 15 minutes. The 15 minutes averaged values are transfered from the computer's RAM to the floppy disc every hour .

Finally a telephone line was erected to connect the W.T. with the diesel-station (fig. 4) .Through this system the WT can be remotely controlled from the diesel-station, which will increase availability, as there will be swift response to any

MYKONOS ISLAND

TABLE I
DATA ACQUISITION EQUIPMENT POWER CONSUMPTION

UNIT	OPERATING TIME	MAX. POWER	ENERGY / DAY
Data Acquisition HP 3421 A	24 h	5 W	120 Wh
Computer HP 71 B	24 h	0,6 W	14,4 Wh
Disc Drive HP 9114 B	0,5 h	6 W	3 Wh
Printer* HP 82162 AB	0 h	11 W	0 Wh

* Only used during service and inspection
THE MAX ENERGY PER DAY IS ~137 Wh

TABLE II
W.T. STOPPAGES

	CAUSES	86	1987												1988		
		D	J	F	M	A	M	J	J	A	S	O	N	D	J	F	M
W.T.G.	NACELLE UNBALANCE	1	1	2	3	3	1		3	4	4	10	3		OUT OF OPERATION		5
	YAW OVERLOAD			1	2									2			
	GENERATOR OVERHEAT		2														
	BRAKE PADS				1	1		1	2	1				2			1
	GENERATOR RPM SENSOR													1			
	WINDVANE											1	1	2			
WIND	AUTO CABLES UNTWIST		1	1	1			1	5	1	4	1					
	EXTREME WIND SPEED	1	2	1	1	2			1	2	3	2	1	2			
GRID	GRID CUT-OFF	3	2	2	2	1	5	1	2	1		3	3	2			
	VOLTAGE TOO HIGH	2		1	8						1	1	1				
	VOLTAGE TOO LOW		1	1	2							1	2	1			
	FREQUENCY ERROR	2	3	2	3	3	3	1	5	1	4	1					

Number of times the W.T.G. has been brought to a stop.

M Y K O N O S   I S L A N D

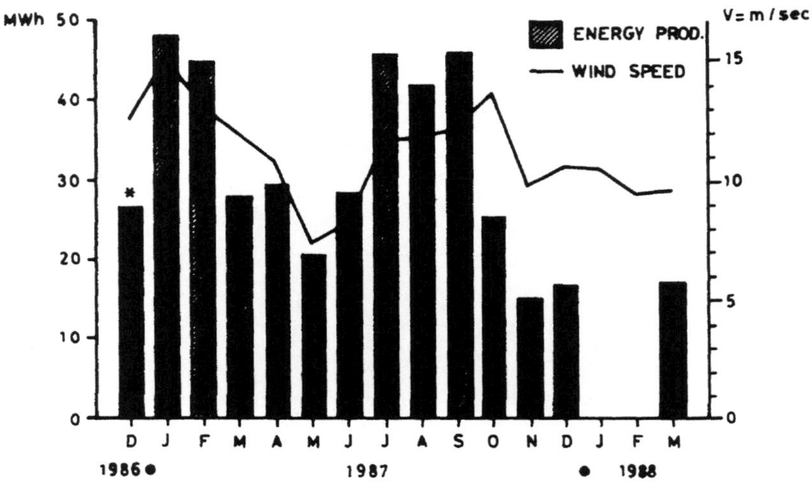

Figure 4. W.T.G. PRODUCTION - MEAN WIND SPEED (at 10 m height)
* START OF OPERATION 8.12.1986

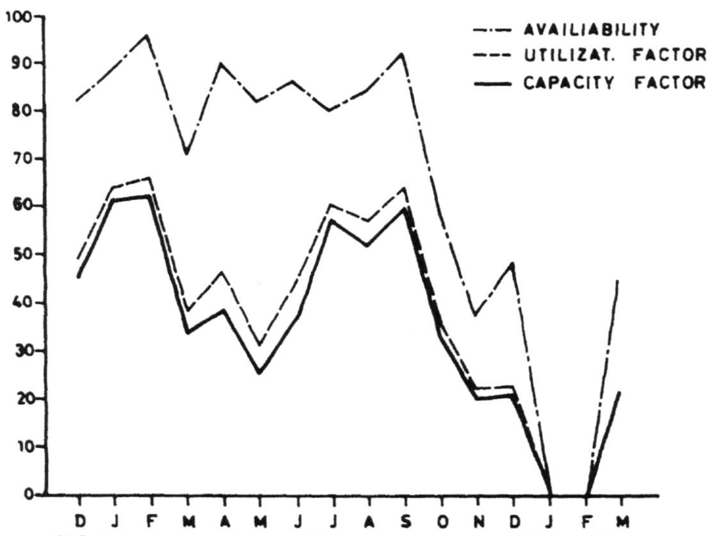

Figure 5. W.T.G. AVAILIABILITY - CAPACITY FACTOR - UTILIZATION FACTOR

failure of the unit, by having the operating condition of the W.T. instantenously displayed at the station.

4. RESULTS OF OPERATION AND MAINTENANCE.

The operation so far can be described as satisfactory in spite of number of problems that appeared in the 15 month operation period.

To start with, the exceptional wind regime in the Fanari site, led to excellent production figures. Thus total production for the period 8.12.86 to 31.3.88 is 436.263 KWh, availability over the same period 64.3% and overall capacity factor of 35.1%. Operation data are presented in figures 4 to 10 and Table II.

If one considere only the period 8.12.86 to 8.12.87, then the results are, total production 410.772 KWh , availability of 74.2% and capacity factor of 43.4%.

It can be seen from the figures that for an interval of almost three months (mid December 1987 to mid March 1988) the W.T has been out of operation. The cause for this was a series of failures that occured one after the other (induction sensor windvane, rpm. sensor , cold soldering) which, taking into account the time lapse before the spare part could be obtained, led to substantial loss of production.

Similarly the months of October and November 1987 have low availability figures which are accounted for by,the frequent occurence of "unbalance" indications, which due to electronics failure could not be reset normally and within a short time interval, leading to production losses. The correction of this fault however was soon followed by the series of failures described above.

The first scheduled maintenance was carried out in August 1987, while the second is expected to be carried out during May.

It is hoped that the installation of the Data Acquisition System and the Remote Control Installation in the diesel station will improve both production and evaluation of operation data.

5. ECONOMIC EVALUATION

The total project cost to date ammounts to over 32.000.000 Drs. which means that actual costs thus far are 10% higher than the project budget estimated in 1984. Furthermore if the expected expences until the end of the demo period are included then the project cost runs to 40.000.000 Drs, an increase of

MYKONOS ISLAND

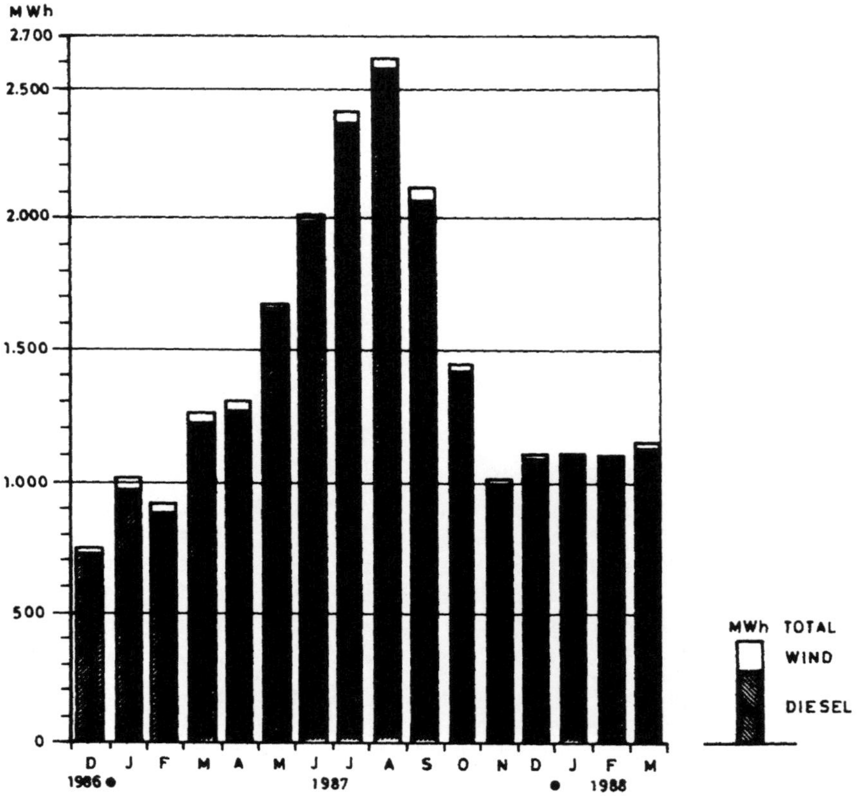

Figure 6. W.T.G.- DIESEL STATION PRODUCTION

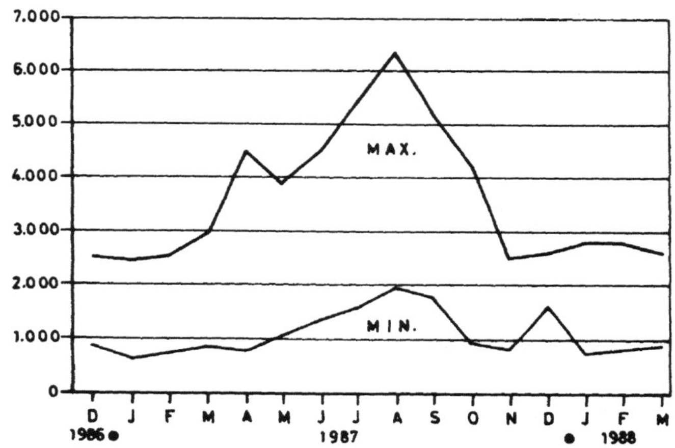

Figure 7. DIESEL STATION MAX-MIN POWER CURVE

MYKONOS ISLAND

Figure 8. WIND ROSE (1983-1987)

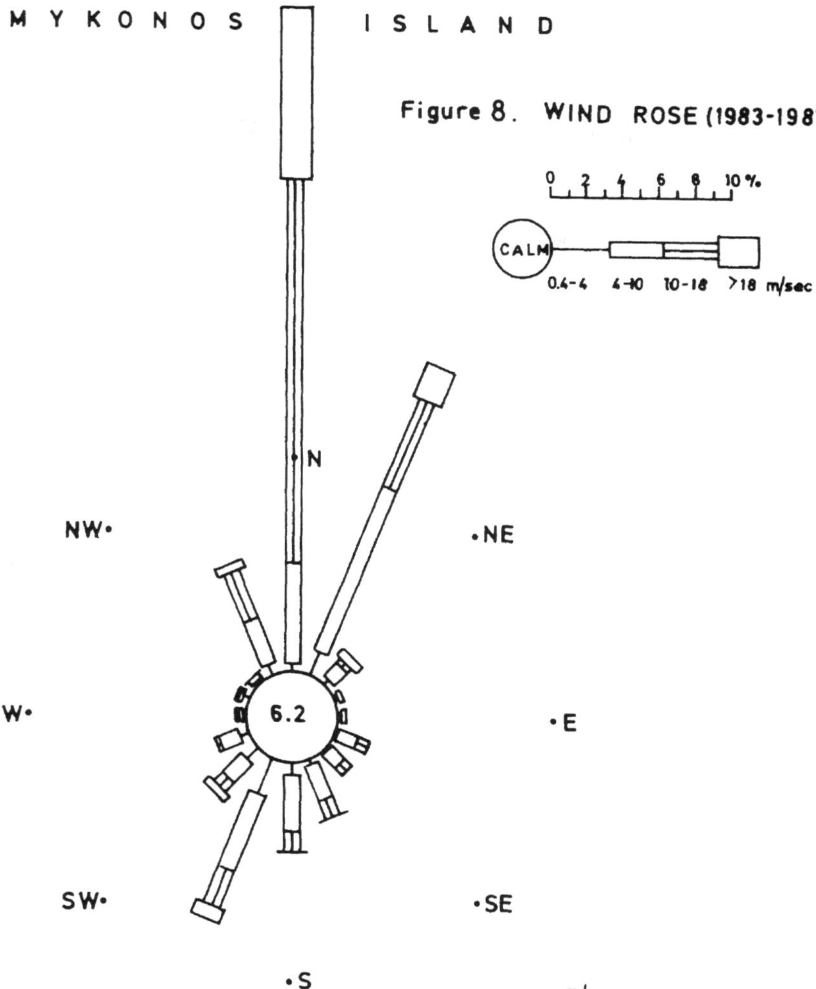

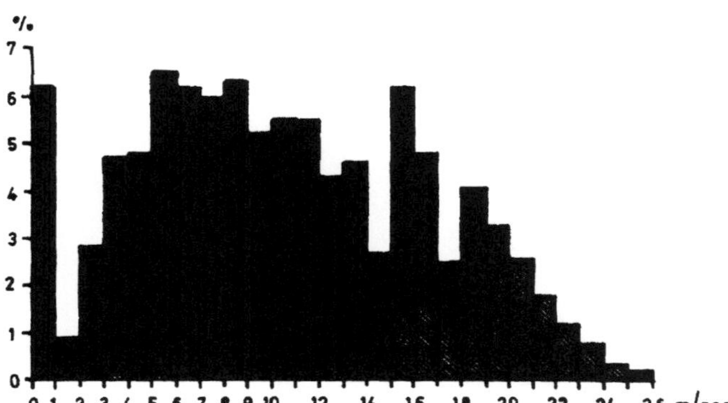

Fig. 9. WIND FREQUENCY CURVE (1983-1987)

(WIND SPEEDS AT 10m HEIGHT)

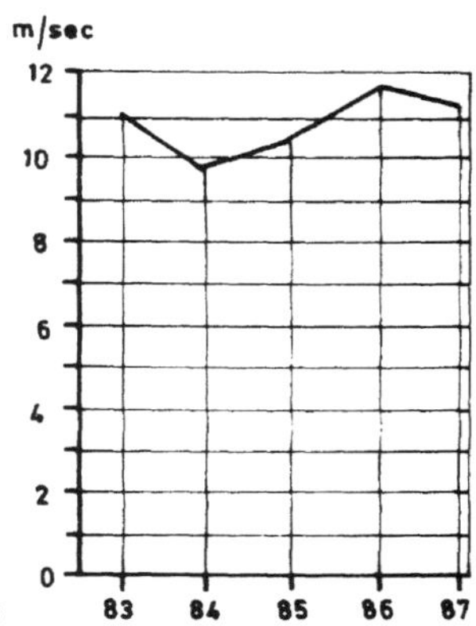

Fig.10. MEAN ANNUAL WIND SPEED.
OVER 1983-1987 PERIOD

approximately 38% on the estimated cost .This increase however can be well accounted for by inflation (1984-88) and the extra costs associated with the project (remote control line etc.)as well as under-estimation of the costs for activities such as transport , erection etc.

To calculate the cost of KWh produced by the W.T. the following formula can be employed:

C= (C/W) (R+M)/h.F

C= Capital invested 30.000.000 Drs. (subtracting the two year demo period cost)

W= Rated power 108KW

R= Annual capital charge rate 10% (20 years life)

M= Operation and maintenance costs 4% (due to lack of industrial infrastructure end access problems)

h= 8760 h

F= Capacity factor = L.A.a

where

L= rated load factor depending on the WT and site wind conditions taken as 0.40

A= availability taken as 0.70 because of the lack of infrastructure and access difficulties

a= efficiency 0.95 (energy losses) and hence

$$G= \frac{30.000.000}{108} (0.1+0.04)/8760X0.4X0.7X0.95 = 16.69 \text{ Drs/KWh}$$

In comparison the KWh cost from the diesel station in Mykonos is 13.00 Drs (1987 prices) . However the high cost of energy produced from the W.T. can be reduced if certain economic parameters associated with this project are not taken into account. Such parameters which are particular to this project are the following:

a) The project is one-off and hence in a series, economies of scale will reduce certain non-reccuring costs per unit.

b) The heavy instrumentation (data acquisition system).

c) The design costs

d) Lack of experience

In such a case the capital invested per unit will be of the order of 20.000.000 Drs. leading to cost per. KWh of 11.13 Drs which compares favorably with the cost of diesel produced KWh.

6.DEGREE OF SUCCESS AND OUTLOOK

For PPC this project can be certainly characterised as sucessful for it gave us the necessary experience in aspects such as foundation, transport, erection and of course operation of a W.T. The fact that the project was completed within its time-schedule and that first year's production surpassed 410.000 KWh is encouraging for PPC, in view of its future wind energy programmes. The experienced gained from this and other demo projects paved the way for the big wind energy programmes such as HORS QUOTA (3.8 MW) M.I.P (4.5 MW) and VALOREN (8.5MW), which will certainly take PPC into the big W.T. league .

Project Nr 619/84 DK

WIND/DIESEL ELECTRICITY SUPPLY, ANHOLT ISLAND

P. Christiansen and E. Damgaard
ELSAMPROJEKT A/S, Denmark

Summary

The electricity supply of the island of Anholt is today based solely on diesel generating equipment. This means that the production price for electricity is comparatively high. The present paper presents a demonstration project where a standard wind turbine is connected to the electricity supply grid of the island. First, it investigates the effect of a wind turbine on the quality of the island's electricity and establishes the expected oil savings.

The investigation shows that a modern 11/55 kW wind turbine can be connected to the grid and contribute with approx. 100 MWh annually without causing appreciable changes of the voltage quality. It is a prerequisite that the connection and disconnection of the turbine is controlled by the production of the diesel equipment, so that the turbine supply maximum 40-50% of the total output. It is further assumed that the the turbine is provided with equipment for a "soft" generator connection to the grid.

Given these conditions, it can be expected that a wind turbine of the size mentioned will make it possible to save approx. 20,000 litres of oil each year.

Assuming an increase of the island's electricity consumption the amount of oil saved will be even bigger, as the above-mentioned restriction of the turbine production will be reduced as consumption increases. The maximum saving will be 25,000 litres annually, when the restriction is dropped.

1. INTRODUCTION

Anholt is a small island with approx. 70 households. It is situated in mid-Cattegat, about 50 km north east of Grenaa. In the summer period, the electricity consumption grows considerably due to the many tourists visiting the island. The load pattern is shown in figure 1.

Today, the electricity supply at Anholt is based solely on diesel generating equipment. The power station owned by the municipality of Grenaa comprises 4 generating sets with a total power of 1143 kVA (915 kW).

A proposal to supplement the electricity supply of the island with a wind turbine has been treated (and passed) by the municipality of Grenaa. The objective of the turbine will be twofold: First, it will save oil and second it will serve as a demonstration project for a wind/diesel system.

The turbine is coupled directly to the grid in a 200 kVA transformer at the harbour of Anholt, see fig. 1. The geographic location of the turbine is shown in fig. 3.

2. CONCLUSION

This study shows that the electricity supply at Anholt may be augmented with a 2-generator wind turbine without noticeably deteriorating the voltage quality.

Load tests on the island have shown voltage fluctuations of an acceptable size. A wind turbine provided with current limiting equipment and a control system which limits the change-over frequency between the 11 kW and the 55 kW generators will cause switching surges up to 2% U_N and with a frequency which will not cause disturbances.

3. ELECTRICITY PRODUCTION OF WIND TURBINE. OIL SAVINGS

In order to evaluate the probable electricity production from a wind turbine, calculations have been carried out according to the rules laid down in RISØs "Wind Atlas of Denmark" (ref. 2). The preliminary conditions are summed up in fig. 3. The calculations were carried out by means of a computer program constructed on above principles.

With the chosen location on the eastern side of the island, a 11/55 kW turbine will be able to generate 150 MWh/year averagely.

The calculated values for the production of the turbine presuppose that the electricity system is able to make use of the maximum effect of the turbine at any time, but due to the small load at certain times of the year (see fig. 2) the turbine will have to stop and start according to the actual load.

Investigations have been made, both regarding a standard 11/55 kW wind turbine, the same turbine geared down to 9/47 kW, and presupposing a maximum load part of the turbine (penetration) of 40% and 50% respectively. Direct tests with an installed turbine will show the highest penetration permissible when the voltage and frequency stability must be maintained.

Load measurings in 1984 and 1985 have led to the conclusion that the wind turbine will have to be stopped for 500 - 1000 hours per year, during which time it would otherwise have been able to generate electricity.

The reduction of the electricity generation will thus be 20 - 60 MWh/year, dependent on the type of turbine and the permissible penetration. This will give a probable generation of approx. 100 MWh/year, corresponding to 10% of the total electricity consumption on the island.

If the electricity production of the turbine is converted into diesel oil saved in the power station, the following figures will be found:

Penetration	47 kW turbine	57 kW turbine
40%	18600 l/year	17300 l/year
50%	22600 l/year	21300 l/year

Attention is drawn to the fact that a 10% reduction of the electricity generation of the diesel generators will only result in an approximate 7% saving of the oil consumption, as the efficiency of the diesel generators is noticeably lower in the actual operating area (see fig. 4).

An increased load may be expected on the island in the years to come, and the saving in oil will thus be increased correspondingly. If a 47 kW turbine is installed, the oil saving will eventually reach 25000 l/year.

It should be noted, however, that the load increase will probably take place by higher load levels, whereas the minimum load does not change to the same extent, since it is determined by the night load in the spring and autumn months and thus by the consumption pattern of the resident inhabitants.

4. VOLTAGE QUALITY

According to calculations of the voltage variations when switching the turbine in and out, the 47 kW turbine will only cause minor disturbances on the grid. The disturbances caused by a 57 kW turbine will be more pronounced but still acceptable.

If the cutting-in of the turbine is controlled like it is normal with modern turbines, the voltage drop will amount to 2% on the connected transformer, since the voltage regulator will have time to compensate the internal voltage drop in the diesel generators.

The mentioned voltage drops will occur by direct cutting-in of the big generator or by switching from the small to the big generator. This change-over may, under rough weather conditions, occur up to 20 times an hour. If they occur "too often", the control of the turbine will interfere and limit the number of change-overs to a few times per hour.

By start-up with the small generator acting as motor and when switching from the big to the small generator, the voltage drops in the connection point will be less than 2%.

When the turbine is switched on causing a switching surge of 70 A this will also cause a frequency swing of up to +/- 2 Hz, but this swing will be levelled in 1.5 - 2 seconds and will thus not cause any inconvenience to the consumers.

REFERENCES

(1) Voltage Quality in 380 V Distribution Network. DEFU Recommendation No. 16.

(2) Wind Atlas for Denmark, RISØ, January 1981, RISØ-R-428.

(3) Summary and Interpretation of some Danish Climate Statistics, RISØ, February 1983, RISØ-R-399, p. 46.

(4) Measured Effect of Wind Generation on the Fuel Consumption of an Isolated Diesel Power System. IEEE Transactions, 1983, PAS-102, p. 1788-1792.

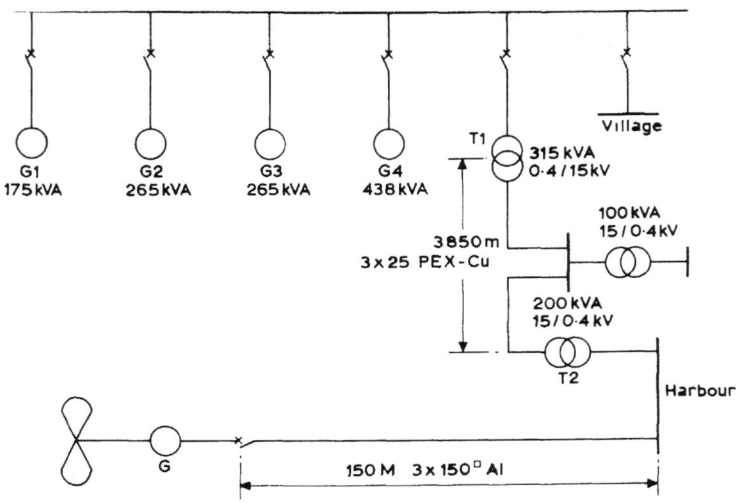

Figure 1 : Electricity Grid on Anholt

ELECTRICAL LOAD CURVES FOR ANHOLT

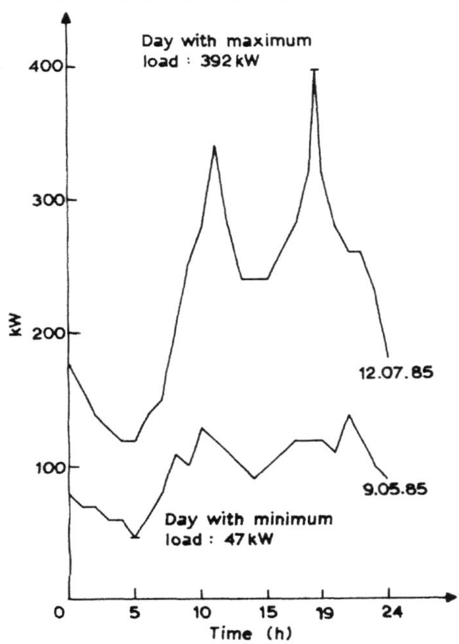

Figure 2

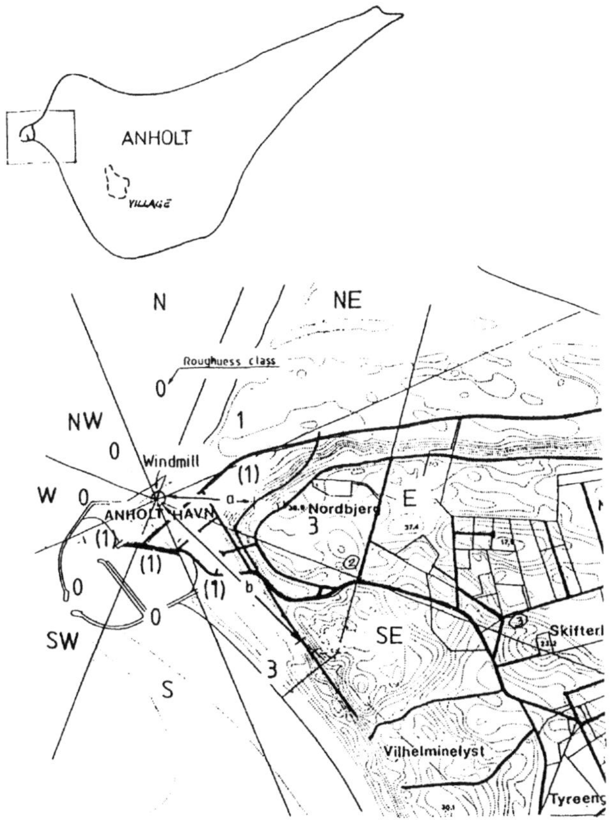

a : 300 m
b : 600 m } Distance to change of roughuess class

Figure 3

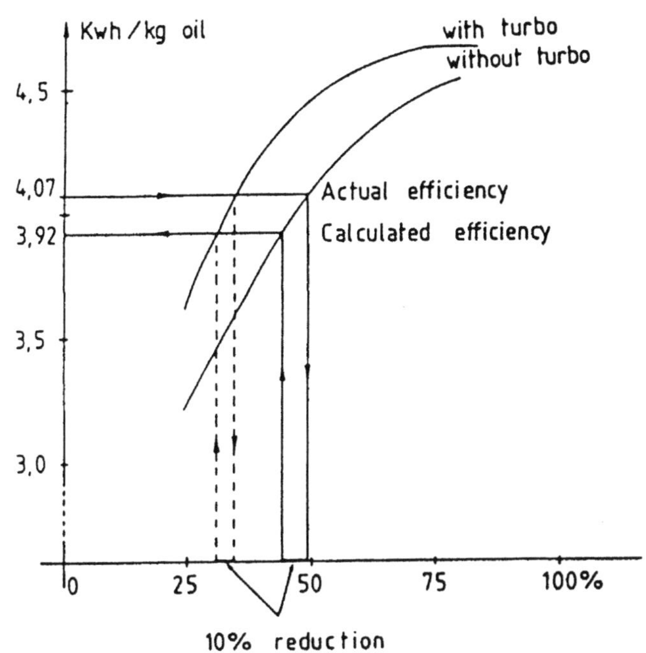

Figure 4 : Efficiency of dieselgenerators

Project Nr 91/85 FR

COMPLEMENTARY ELECTRICITY AT AMSTERDAM ISLAND WITH A VAWT DARRIEUS TYPE, 10 M DIAMETER

P. PERROUD, G. BERTRAND and X. PLANTEVIN
Commissariat à l'Energie Atomique - 85 X - 38041 Grenoble Cedex (France)

B. BONNET
Terres Australes et Antarctiques Françaises,
34, rue des Renaudes, 75017 Paris (France)

Summary

The objective is to show the interest of utilizing wind energy in an isolated site in order to save fuel. Amsterdam island, located South of Indian Ocean, at 2 880 km from La Réunion island, presents the required conditions for this purpose: good wind resources, diesel plant, technical people of TAAF in place.
Main characteristics of the machine: 3-bladed, solidity 0.18, constant speed 93 rpm, rated electrical power 30 kW at 13.5 m/s can reach 36 kW for higher wind speeds. Carried by ship "Marion-Dufresne" the wind mill D10-2 was erected at Martin de Viviès base by a joint team CEA-TAAF in fall 1986 and then coupled to the diesel plant Dec. 19th 1986. It has been working during 1987 under supervision at a reduced power, because of a deficiency in the braking system. A new braking device redundant and more powerful, including emergency brake has been installed Dec. 1987 in place of the previous one. Thus, experimentation will proceed normally. Yet, fuel saving is of the order of few percent but, it is expected to reach 17% of the yearly consumption.

1. INTRODUCTION

The objective is to show the interest of utilizing wind energy on a remote site, in view of fuel saving. Amsterdam island, located in the southern part of the Indian Ocean at 2 880 km ESE of La Réunion island nearby Madagascar, was chosen because this site has the required conditions: good wind availability, autonomous diesel plant, technical people of the yearly TAAF expedition of about thirty men. The diesel plant is equiped with three engines no-connected in between: installed power 120 kW.

The Amsterdam island of French sovereignty, depends on the Administration of Terres Australes et Antarctiques Françaises (TAAF), organism related to the Ministery of Territoires et Départements d'Outre-Mer (DOM-TOM). TAAF includes four districts: Amsterdam, Crozet, Kerguélen and Terre Adélie (Antarctic continent) (Fig.1). Its weather is relatively mild, tempered by the ocean.

The VAWT D10-2 installed at Amsterdam island (Martin de Viviès base, Longitude 77°32'E, Latitude 37°47'S) is a Darrieus type, 3-bladed, rated electrical power 30 kW, designed and constructed by the Centre d'Etudes

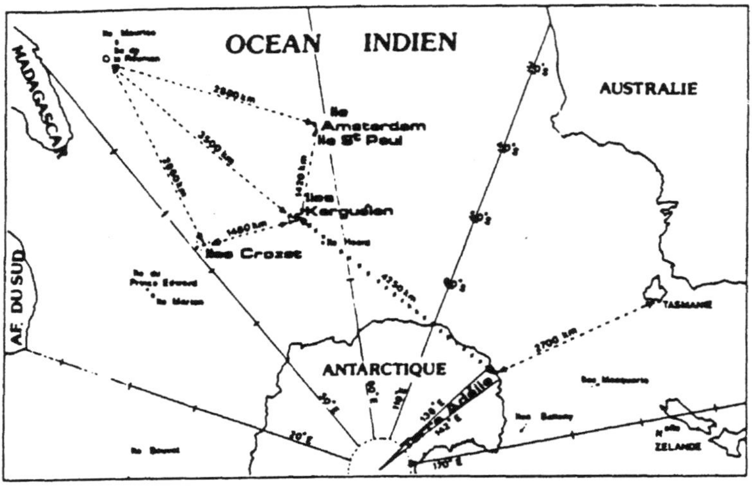

Figure 1 : General map of the South Indian Ocean showing relative locatio of the four TAAF's districts

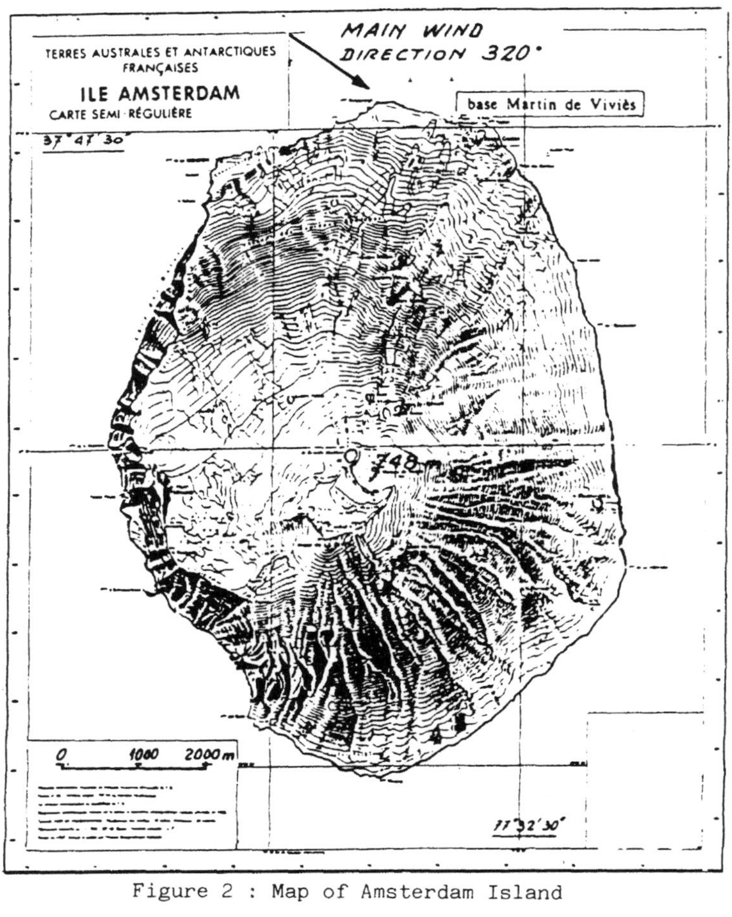

Figure 2 : Map of Amsterdam Island

Nucléaires de Grenoble. It is the second of this kind, the first one, D10-1, was installed at CENG site in April 1984, then transported at the Lastours (Aude) site in August 1986. These two wind mills are identical and manufacturing is particularly good. Both of them operate at constant rotational speed, connected either to a large grid such as E.D.F. network, or to a diesel power plant (1, 2, 3).

For wind speeds ranging from 12 to 25 m/s, a daily energy production of 400 kWh has been recorded, thus demonstrating the good performances of Darrieus rotor. The Amsterdam machine was conceived to operate under sea climate conditions.

Transported by the ship "Marion-Dufresne", the D10-2 was installed by a joint team CEA-TAAF, December 1986. It has been since operated during year 1987 under control and at reduced power, because the lack of reliability of the braking system. A new braking device, redundant and overpowerful including emergency brake, has been installed at the place of the former one in December 1987, allowing thus normal operations (Photos 1 - 2)

At first, a yearly production of 71 000 kWh was estimated leading to a fuel saving of 26 m^3, i.e. 17.4% of the yearly consumption of 149 m^3. That is the goal to be reached!

2. MAIN CHARACTERISTICS OF AEROGENERATOR D10-2

Vertical Axis Wind Turbine, Darrieus type. Rated power: 30 kW at V = 13.5 m/s. Maximum power: 36 kW at V > 20 m/s.

2.1. Rotor

3-bladed (n=3).
Height = equatorial diameter = 10 m, H/D = 1.
Swept area: S = 67.7 m^2
: $\sigma = nC/R = 0.18$, R = D/2.

Blades : extruded aluminium 6060.CEGEDUR-PECHINEY.
Symetrical NACA 0015 profile, chord C = 300 mm, maximal thickness 45 mm, linear mass 8.1 kg/m, metal area 3000 mm^2, Young modulus 69 000 MPa.
Troposkian shape (parabola and straight ligne), length 13.655 m.
Binding blade/rotating shaft: articulated on a steel axle Ø 60 mm with the help of two bolted half shells.
Center of mass at 3.170 m apart from the vertical axis.

Rotating shaft:
steel tube, 610 mm diameter, thickness 6.35 m in two parts assembled with bolts, linear mass 94.5 kg/m.
Helicoïdal ladder welded on the tube.

Bearings : Ball and socked joint, waterproof, lubricated with grease.

Height of the rotor equator: 12 m.

2.2. Tower

Steel tubular tripod, 6 m high, can be dismantled.

2.3. Guy wires

Three stainles steel cables, 22.2 mm diameter, length 29m, linear mass 2.16 kg/m.

2.4. Power train

. Automatic safety brake on the low speed shaft. Disc diameter 900 mm.

. Elastic coupling.
. Gear box: planar, two stages, ratio 16.1/1.
. Two induction generators, 1500 rpm, 380 V, 50 Hz.
. Power 11 and 18.5 kW. Mounted on the same shaft, coupled in parallel on the grid. Constant speed operation 93 rpm. Generator 11 kW is used temporarly as a starting motor. According to electrical power, either one generator or both are connected to the grid.
. Speed control by tachymetric alternator placed at shaft end. Automatic operation by means of an electronic control unit monitoring: starting, coupling, de-coupling, stopping, measurements, safety, etc.
Cut-in 6.5 m/s. Cut-out 20 m/s. Maximum design 60 m/s.
. Rated power 30 kW at 13.5 m/s.

2.5. Electronic regulator of power load

In case of diesel connection, this auxiliary device ensures a good electrical stability, the excess power produced being dissipated in a dump load. This dump load consists of 12 single phase 2.5 kW resistors, which are switched on by an electronic regulator based upon semi-conductor rectifiers (1,2,3).

2.6. Remarks on the safety brake

It has been reinforced since 1986 and includes now a disc, 980 mm in diameter, thickness 27 mm instead of 12.7 mm, equiped with three shoes-clips controled independently: 2 electrical, 1 air pressure (emergency brake). This redundant and powerful system has a braketorque of 20 640 mN instead of 5 400 mN as formerly. The machine can be stopped within 10 seconds.

2.7. Mass distribution

Rotor	: 3 186 kg
Power train	: 1 091 kg
Tower	: 3 579 kg
Guy wires	: 525 kg
Total	: 8 381 kg

2.8. Corrosion protection

Zinc and polyurethane coatings. Stainless steel. Grease.

2.9. Lighting protection

Grounding with flexible cupper conductors for the three guy wires, as well as for other points.

3. THE PROJECT

It includes five phases:
Phase 1: Manufacturing and mock-up in France.
Phase 2: Preparation of the Amsterdam site.
Phase 3: Dismantling, packing, transportation, landing, installation, electrical connections.
Phase 4: Putting into operation.
Phase 5: Exploitation. Technical assistance. Maintenance. Results.

As soon as notification of the contract by CCE Nov. 27th, 1985, the project was launched. It has been conducted by Mr Paul PERROUD, at this time head of Lab. ASP at CEA-CENG, in connection with Mr.Bernard BONNET, head of Technical Services of TAAF in Paris.

The project realisation matched exactly the initial planning because a rigid time schedule: ship "Marion-Dufresne" time-table either at

Marseille or at La Réunion harbors. Missing one of the dates would have resulted a one-year delay. The installation at Amsterdam can be done only at the changing of the team, i.e. in November-December. Therefore, a strict planning of the different phases was set-up. This intense work, releasing energy, appeared fruitful. Someone said: "You are sentenced to succeed..."

The VAWT D10-2 produced its first kilowatts Dec. 1986. During 1987, it runned under supervision and at reduced power because lack of brakes reliability as said. The order was to stop the machine for winds higher than 15 m/s.

The exploitation of the Amsterdam wind mill is entrusted to the men in charge of the diesel plant who were trained before departure on the Lastours machine (D10-1) (1).

The overall responsability is taken by Bernard BONNET of TAAF and technical assistance by Lab. ASP at CEA-CENG,whose head is now Mr T.ALLEAU.

Communications with Martin de Viviès base are possible either by telex or by telephone via satellite. Ships seldom visit Amsterdam.

A short report on the wind mill operation is sent in France each month. At this time, because of some disturbances, D10-2 power contribution is rather modest, few % only. We are expecting a more important production in 1988 and after.

Chronology of the operations:

April 1985 : The project is submitted to CCE/DG XVII.

Nov. 27th : Contract notification by CCE N° WE/091/85FR.

Dec. : Preliminary studies according to cranes disposal at Amsterdam.

Jan.-Fev. 86 : Project design.

Fev.-Aug. : Drawings. Selection of sub-contractors.

Mai 15th :Sending to Amsterdam by ship "Marion-Dufresne" from Marseille, stainles steel bars to be put in the concrete anchors.

June 15th- Oct. 15th : Amsterdam site. Construction of the road (60 m) and the plat form (Ø 28.5 m). Altitude 40 m. Distance to the plant 100 m. Basaltic lava excavation 7500 m^3 . Concrete anchors 110 t, iron 2.2 t. Low density of material 1.7 t/m^3 .

Sept.-Oct. : CENG site. D10-2 mock - up. First runs. Packing. Transportation to Marseille by trucks. 12 casings. Total weight 15 t (large number of spare parts). Charging on board "Marion-Dufresne" Oct. 10th. Ship departure Oct. 11th. The electrical control-command unit will join the boat by air at La Réunion with MMr G. BERTRAND and X. PLANTEVIN of CENG/ASP.

Nov. 27th : The "Marion-Dufresne" arrives at Amsterdam after a tour at Crozet and Kerguélen. Landing of casings with towed-air-rafts, the ship being anchored at 800 m from the coast. An helicopter is also used.

Dec. 1st-23rd: Erection of D10-2. Crane, tractor, truck... Blades bending. Construction of a concrete shelter 3 x 3 m. Electrical connection to the diesel plant (5 conductors, 25 mm^2 each, 100 m long). Adjustements. Tests. Production of the first kilowatts Dec.19th.

Dec. 23rd : Embarking of G. BERTRAND, X. PLANTEVIN and B. BONNET on "Marion-Dufresne". Return to France. A. LEROY and R. FAURE in charge of D10-2.

Photo 1 : VAWT Darrieus type D10-2 installed at Amsterdam island (TAAF). Jan. 1987. In the foreground is "Ventôse" station for wind speed measurements. The anemometer is at 12 m above ground level, i.e. rotor equator height. The control-unit-system is placed in concrete shelter.

Photo 2 : Amsterdam island (TAAF). General view of Martin de Viviès base. Jan. 1987. In the foreground is the VAWT D10-2. Notice the fence for wild cows protection.

Dec. 5th 86-Feb. 28th. 87 : D. PONCET's trip, of Alpes-Automatic Co. A second control-command unit is transported. At Amsterdam: electrical tests of coupling.

Nov. 5th 87-Feb. 29th 88: X. PLANTEVIN's trip. Modification of the braking system.

4. COMMISSIONING

Project management: CEA-CENG/ASP. Design. Drawings. Sub-contrators selection. Assembly of the parts. Mock-up. Tests. Sub-contractors were found among local contractors.

Main sub-contractors

. Blades: CEGEDUR-PECHINEY at Issoire (Puy de Dôme).
. Blades bindings, axles, etc.: Ets ANDRE, St Martin d'Hères (Isère).
. Rotating shaft: Ets ANDRE, St Martin d'Hères (Isère).
. Brake disc: Ateliers LERAT, Grenoble.
. Brake shoes-clips: TWIFLEX-UNICUM,
and Ets POMMIER, La Tour du Pin (Isère).
. Tower: Ets BRUN, Fontaine (Isère).
. Electrical power train: LEROY-SOMER, Angoulême (Charente).
. Control-command unit: ALPES-AUTOMATIC, Le Versoud (Isère).
. Guy wires: SARMA, St Vallier-sur-Rhône (Drôme).
. Coating protection: EGPS, Claix (Isère).
. Packaging: PRESTEMBAL, Saint-Egrève (Isère).
. The mock-up at CENG site was carried out with the help of "M-Industries", MONTAZ Brothers, few days before departure.

Blades, equipped with their bindings, were not bended for transportation easiness. Bending was done on the spot with a special tool.

This preparation work could be achieved in a limited time (3 weeks), thanks to the good will of all the partners.

5. RESULTS

On Fig. 3, 4, 5 and 6 is shown the effective electrical power versus wind speed P(V), as recorded on our three sites. From 12 m/s, the mean experimental curve leaves the theoretical curve computed with IREQ code, to reach a ceiling at 20 m/s, i.e. at $R\omega/V = 2.5$. Thus, the VAWT power does not decrease as predicted by theory for higher wind speeds. Ignoring this fact could leads to some troubles...

In our case, it can be seen that the real maximum power exceeds the rated power by 20%. It may result an over-heating of the generators and a safety stopping of the machine. It is therefore more careful to stop automatically the wind mill at 30 kW by acting on the out-put relay. Thus, some energy is lost, but wind availability above 20 m/s is rather small at Amsterdam. It is concluded that efficiency of Darrieus rotor overpass the predictions ... and the choice of D10 rated power was somewhat underestimated. A generator of 40 kW would better fit this machine.

Since its coupling to the diesel plant, end of 86, the D10-2 has runned a little part of the time, because it was limited to winds smaller than 15 m/s. It was stopped during night time. Moreover, the needs of the base are for the time being, smaller than those estimated, and the adjustements of the installation took some time. Hence, the number of 17%, expected as fuel saving, is not yet reached.

During the year 1987, this number was of the order of few percent only. The results of this year is not representative of normal operation of our VAWT.

By utilizing an auxiliary device, whose function is to adjust instantaneous power production to power needs by driving the excess electricity to resistors dump load (§ 2.5), it will be possible to let the wind mill in continuous operation. But, this technique, is an energy waste!

A better solution could be found in storing wind electricity produced by heating water for Martin de Viviès base. This project is being studied by TAAF/ST.

Each month, the man in charge of the D10-2 at Amsterdam sends to TAAF-Paris information on the operations: number of hours of run, wind availability from "Ventôse", energy produced by the diesel plant and by the wind mill, technical problems encountered, etc.

6. TIME-TABLE

Finally, the initial planning of the project written in the contract has been followed according to TAAF's ship "Marion-Dufresne" cruise dates.

7. COSTS

Total cost including the five different phases (§ 3) exceeds the initial estimates by a factor of 20%. Part of the extra-cost is due to spare parts and braking system modifications.

We don't have, at the present time, enough hours of operation in order to determine precisely the cost per produced unit of energy. This could be done by the end of 1988.

It is obvious that the installation of such machine in a remote site far away from France, such as Amsterdam island, necessitates costly investments, more costly than conventional means of energy production. It is a "première" for us. However, a second wind mill of the same type, taking advantage of the experience gained, would be less expensive. If the cost of the materials, the manpower, and transportation are well known in France, it is not the same overseas.

The comparative cost of an eolian system, which is complementary with a conventional system, is rather difficult to make. Truly speaking, it would be better to say: "I utilize the winds of Antartica..." an unexhaustible energy source, rather to claim pointless speculations. Nevertheless, each ton of saved fuel, is one gained ton. It is less air pollution. It is also a good contentment, keeping in memory that the results over a long period of time (10 years) would consolidate this intuitive philosophy of the moment.

8. PRESENT STATE AND OUTLOOK

To our opinion, this achievement has been well carried-out: the VAWT D10-2 works in Amsterdam island and it is in the process, of improving its production.

The other islands of the TAAF: Kerguélen, Crozet, which even have a better wind availability, could be equipped as well in wind mills giving complementary electricity to the bases. Which type of machines to select? We have thought, this is normal, that the Darrieus rotor, though somewhat more expensive, would fit better the situation than HAWT machines because of its greater robustness. Among the French constructors which one to select? What is the present and the future of our wind energy programme? Questions without clear answers! Maybe, we will have to open this market to some of our partners of the EC, more advanced than our country in this field? The Europe of 1993 will, perhaps, do the job!

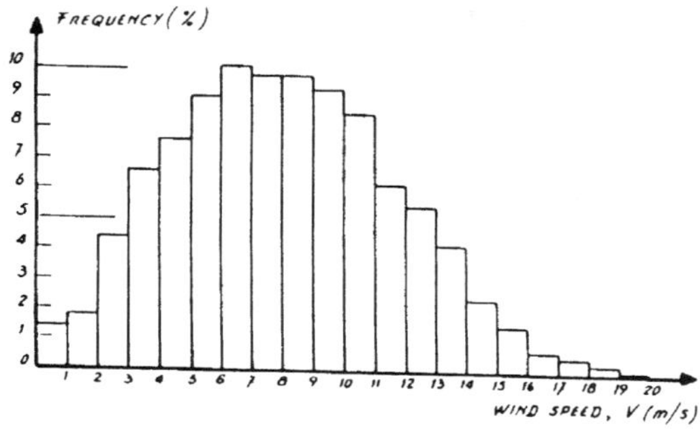

Fig. 3 : Amsterdam Island. Wind speed availability.
Height measurement : 12 m.
Period: 1961-1970
Main wind direction : 320° (NW)

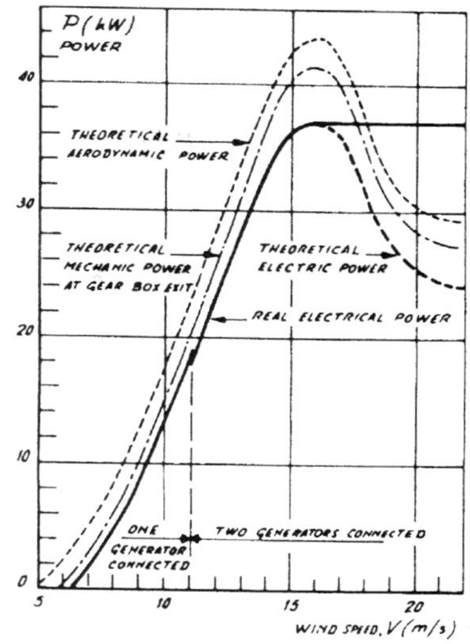

Fig. 4 : <u>VAWT DARRIEUS TYPE D10</u>
Theoretical and real performances.
Aerodynamical power was computed from IREQ's code. Mechanical and electrical power are deduced taking into account from the drive train efficiency

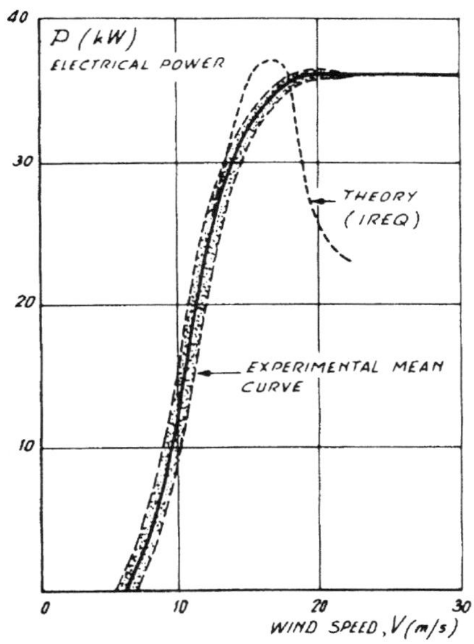

Fig. 5 : Aerogenerator D10.
Effective electrical power

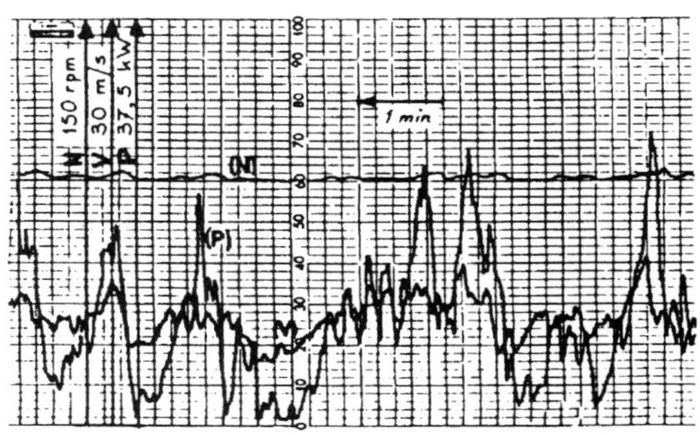

Fig. 6 : Example of recordings at Amsterdam.
V (Wind speed),
N (Rotational speed),
P (Electrical power).
Jan. 10th, 1987

9. INDUSTRIAL DEVELOPMENT AND MARKET EXPLOITATION

The Commissariat à l'Energie Atomique having achieved its own wind energy programme (over 10 years) because of the tightening of its activities, will not go further in the development of Darrieus rotors towards greater power (CEA/NEYRPIC project of a 350 kW wind mill was canceled in 1986). The two D10 yet constructed, are advanced prototypes, ready to go to industry; there exists also very complete detailed drawings and serious experienced men, having taken part to the different operations. In fact, a wide know-how is still available.

We are waiting for a clever contractor, French or from the CCE, who would go ahead. But, this depends on the world market.

In France, nuclear energy supplies 75% of the electricity consumption with a relatively low kilowatt cost, and E.D.F. grid covers all parts of the country. Abroad, there exists the DOM-TOM energy needs.

NOTE

In the course of the different phases, this project showed a continuous interest. It has exalted those who believed in its success because it was, in addition, a piece of Adventure.

ACKNOWLEDGMENT

This project was supported by: CCE (DG XVII), CEA (IVI), TAAF and AFME (DOM-TOM).

REFERENCES

(1) PERROUD, P., BERTRAND, G., SALANON, J.M. and TERRIER, G., Development of vertical axis wind turbines, Darrieus type at Grenoble.EWEC 1984. Hambourg, RFA, Oct. 1984.

(2) PERROUD, P., BERTRAND, G. and SALANON, J.M., Les éoliennes type Darrieus de Grenoble. Les vibrations et les transitions électriques. Communication présentée au Séminaire International sur le comportement dynamique des éoliennes. Sophia-Antipolis, 28-30 Avril 1986.

(3) PERROUD, P., BERTRAND, G. and SALANON, J.M., Darrieus wind turbine, 10 m diameter, power 30 kWe. Connection to a power grid or to a diesel plant. Communication présentée à EWEC 1986. Rome, Italia, 1986.

Project Nr 376/86 ES

"HYBRID WIND-DIESEL SYSTEM FOR COMMERCIAL EXPLOITATION"

J.P. TORTELLA
Gas y Electricidad, S.A.
Dpt°. Planificacion C/D y Nuevas Actividades
C/.Juan Maragall, 16-6a Planta
07006 PALMA DE MALLORCA - SPAIN

1. INTRODUCTION

1.1. Project Description

A 10 m. Diameter, 30 kW horizontal axis wind turbine, model GESA-10 ; designed, developped and constructed by GAS Y ELECTRICIDAD, S.A. (GESA) will be installed at Wind test field located in ALFABIA (Mallorca) and will be combined to a 30 kW diesel generator forming an autonomous system. The energy generated will feed a small network of users located 2,5 Km or more from the main grid and not connected to it. The wind turbine has an induction generator which is connected to the rotor through a parallel axis gear box. Operational wind speeds of the wind turbine are between 4 and 25 m/s. The blades are made of glassfiber and are fixed on the main shaft. The nacelle is located on the top of a 12 m. high tubular tower. An active yawing system aligns the wind turbine with the average wind direction.

An appropriate designed central system will regulate the whole wind-diesel system in order to achieve maximum reliability and optimal operation. The estimated annual energy yield by the wind turbine is 55 Mwh/yr. The payback time of the demo project is estimated to about 10 years while commercialization could bring down this figure to 5-6 years.

1.2. Aim/objective of Project

The aim is to reduce the investment cost and improve the economical viability of decentralised electricity production by means of a hybrid wind-diesel system in order to obtain valuable data on reliability, efficiency and economy. It can be achieved by optimizing the control system for stand-alone applications.

The goal of this demonstration project is the basic development and testing of a hybrid wind-diesel system for commercial exploitation. A load management system will set priorities, voltage and frequency supplied to the loads and the electrical resistors buffer storage. Thus the wind generated electricity will be fully used. A global control system will control the wind-turbine, diesel system and load management control subsystem. This unit will receive all variable parameters and emergency signals or conditions and will issue the proper orders to the different control subsystems.

1.3. Site of the project

During 1984 the installation of the "ALFABIA" test station for small WECS, at the Island of Mallorca (Spain), will be completed.

Wind data, measured beside the test station during the seventies, which demonstration the good conditions to test wind machines under normal and extreme wind conditions (gusts of 50 m/s are measured almost every year) are presented.

A description of the test site and of the data acquisition system, allowing the fulfillment of power performance and intensive test, are also presented.

The test station is located in the central part of the Sierra of Tramontana which is basically formed by a wide strip, of about hundred kilometres in length by fifteen in width, which runs parallel with the NW, slope of the Island of Mallorca. The test station is located at 1000 m. of height and access to tit, from Palma, is achieved by a road of 30 km., paved all the way.

As regards the winds, the Sierra of Tramontana acts as a perturbing element, especially in the case of those which come from a N-NW direction, that are channelled by the valley of Soller towards the area of the test station. When the wind direction is parallel to the Sierra the speed at the crest and on the plains practically coincide. When the direction is perpendicular, the ratio between the speed on the crest and the plains is higher than 3.

Beside the test station, the Spanish National Institute of Meteorology, registered wind data during the seventies. Summaries of these data were obtained for most of the years and hourly data, from strip charts, were found only for the year 1973.

In table 1 maximum values of wind speeds are presented and, in table 2, the number of days during which the wind reaches speeds higher than 90 Km/h.

YEAR	MONTH												ANNUAL
	I	II	III	IV	V	VI	VII	VIII	IX	X	XI	XII	
1971	-	178	128	128	90	84	88	84	96	118	168	120	178
1972	148	180	170	164	94	108	130	104	120	156	133	146	180
1973	160	186	136	-	96	126	108	102	120	110	130	160	186
1974	184	200	176	122	158	122	122	78	167	208	-	-	208
1975	-	-	192	146	104	80	-	-	94	140	-	-	192
1976	151	127	151	158	83	70	100	73	112	166	162	173	173
1977	155	108	137	147	142	96	99	124	144	107	157	124	157
1978	171	140	171	196	139	101	-	135	121	165	157	187	187
1979	178	169	185	124	75	101	110	124	103	96	158	189	189

TABLE 1 : Maximum values of wind speed (km/h)

YEAR	MONTH												ANNUAL
	I	II	III	IV	V	VI	VII	VIII	IX	X	XI	XII	
1976	6	6	4	6	0	0	2	0	4	9	8	8	53
1977	5	3	2	4	2	2	1	3	1	4	9	6	42
1978	15	14	13	9	5	3	0	2	3	5	5	16	90
1979	12	17	11	5	0	1	1	4	2	3			

TABLE 2 : No. of days with winds higher than 90 km/h

On figure 1 it is possible to see the main elements of the test station. The control and measuring building is conveniently insulated, from the thermal and electric point of view.

The two foundations, adaptable to different types of towers, dispose of appropriate pieces of masonry for the location of the capstan of the hoisting system, provided in order to permit the installation of the machines without needing to use cranes.

The three meteorological masts have a height between 20 and 30 meters, due to the irregularity of the land, and they practically surround the two machines that can be simultaneously tested.

The foundations and the masts are communicated with the house by accessible above ground channels for the protection of connecting and measuring cables.

READING

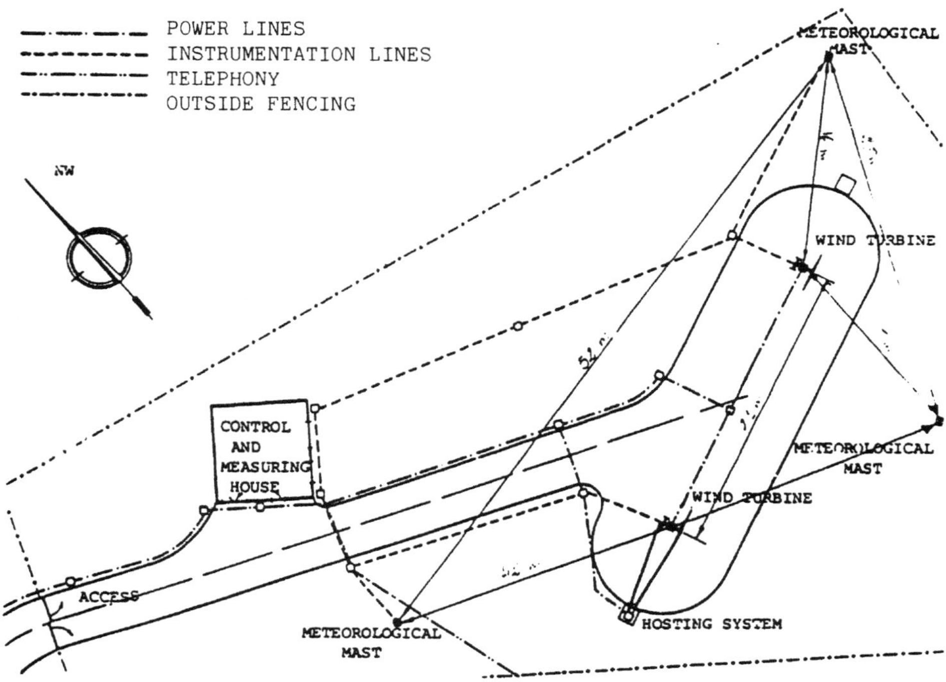

Figure 1 : Map of the "ALFABIA" test station

On the meteorological masts the following sensors are located :

- 9 cup anemometers (3 per mast).
- 3 vanes (1 per mast).
- 2 axial flow anemometers.

For the signals coming from the nacelle, power, tension, current and frequency transducers, and temperature sensors are provided.

The characteristics of the test site and the data acquisition system will permit to test all aspects of a WECS ranging from : power performance, quality of power, fatigue evaluation, noise characteristics, reliability, durability, safety conditions, etc.

It will be intended to follow the Recommended Practices for Wind Turbine Testing and Evaluation of the International Energy Agency and their rules for approval of windmills in Spain, under preparation by the Ministry of Industry and Energy.

2. MAIN DESCRIPTION OF THE HYBRID WIND-DIESEL SYSTEM

The 30 kW Wind-Diesel system can be characterized as a system without electrical energy storage, but we can confirm the interest to include a thermal dump load (resistors) in order to assure the necessary regulation of the system. Some main data for the system are :

2.1. Wind-turbine

Described in point 1.1. and shown in figure 4.

2.2. Diesel generator set

- 4 Cylinders diesel, 48 CV (DIN 6270-A) at 1,500 rmp., water cooled and constant speed centrifugal regulator.
- 31.2 kW (39 kVA), 4 poled synchronous alternator and cell-excited type. (± 1,5 % Nominal Voltage).
- Diesel and alternator connected via an electromagnetic clutch.
- Electrical automatic start.
- Consumption : 172 gr/cv/hour at full power (gas-oil 10500 Kcal.h.)

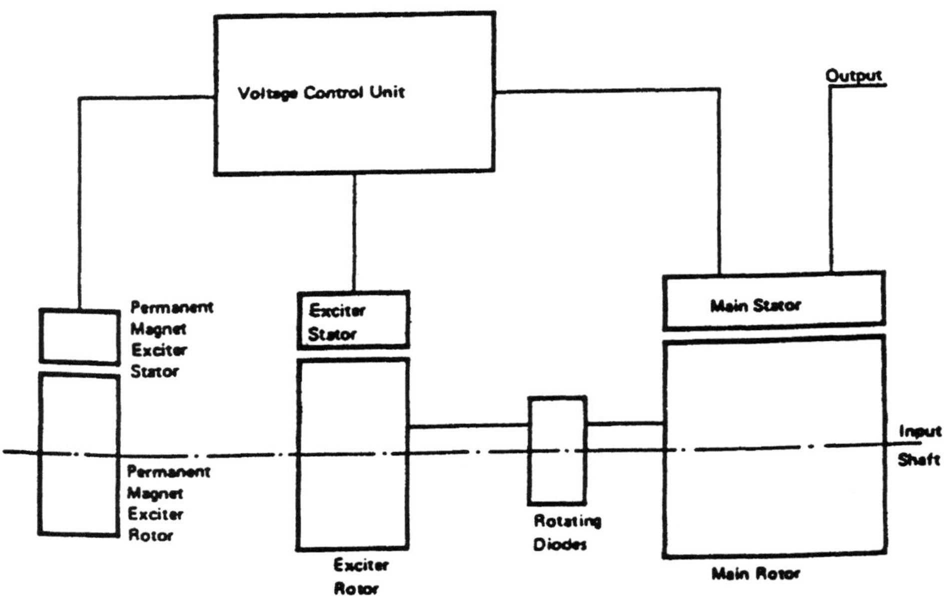

Figure 2 : Block Diagram of Excitation System

2.3. Dump-load

A 15 kW resistor dump load which through switches it can be regulated in steps of 1 kW.

2.4. Control system

A microcomputer based system to control the wind-turbine, the diesel start-stop, the coupling/uncoupling of the synchronous alternator from the diesel, and the dump load.

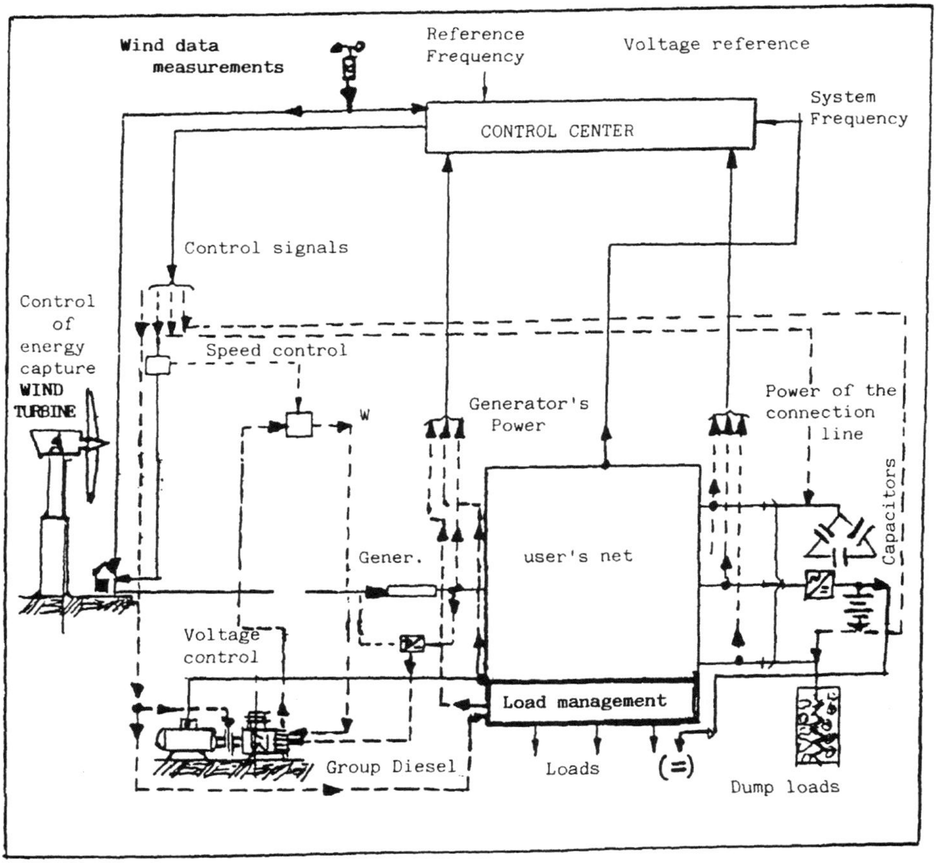

Figure 3 : Block diagram of the wind/diesel system

3. PROJECT REALISATION AND CONSTRUCTION

3.1. Tasks were carried out

The status of the project is that we have been working on the first phase concerning the design and construction of the complete system. We have finished with a collaboration project with a Balearic Islands University, the subtask concerning to the modelisation of the system as a research tool for the examination of wind-diesel behaviour in order to provide a rapid and inexpensive detailed information on the static (step by step) and dynamic behaviour of these systems. The model identifies the direction and magnitude of total system performance changes caused by variations within the subsystems, and permits an examination of changes in system performances from variation in one or several elements.

* Inputs : Wind speed, load.
* Design parameters : 25 (Generators characteristics).
* Common variables : 20.
* Outputs : rotational speed, torque, current, electrical power, cosy, frequency, mechanical power, load power.

3.2. Timetable

The tasks that are to be completed, appreciably differ from the planned scheduled including at last report. At this moment, the delay over scheduled works could be 2 months owing to changes in the structure of our company, with changing of our director and stopping the works temporarily in order to study the convenience of each project with the head project.

The scheduled works are the following :

Project : WE/376/86-ES

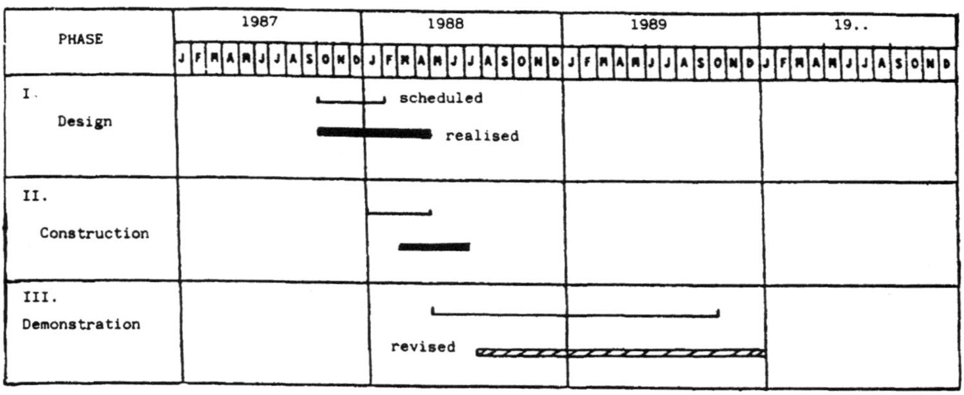

TABLE 3 : Projects' time planning

4. CONCLUSION

The project comprises design, manufacture, installation and performance monitoring of an electricity generating unit consisting of a wind turbine asynchronous generator and a diesel generator of the same output. Aim of the project is to supply on a continuous basis electricity to a remote settlement not connected to the public electricity network and is divided into 3 main phases including :

I. Design of the hybrid Wind-Diesel system.
II. Procurement, W-D system construction, testing construction, testing (on bed and field) and installation.
III. Performance monitoring and analysis to aims its future optimisation and commercial exploitation.

The value of information obtained by use of the computer model is a direct function of the accuracy of the model. While the model appears to provide reasonable results a validation of the model with detailed observations from one wind-diesel system appears necessary.

Progress has been made during the last six month on the different subjects within the Gas y Electricidad, S.A. Wind-diesel programme, in spite of a 2 month delay in the works.

The system has being designed to run in either of the three normal modes :

1) Wind turbine alone (synchronous alternator as reactive energy generator-clutch uncoupled).
2) Diesel and synchronous generator alone (clutch coupled).
3) Diesel synchronous generator and wind turbine in parallel.

In the design of the control system, importance has been attached to the permitted running range with regard to adjustment of the control signals to the dump load and start/stop of the diesel. Furthermore, we are trying to maintain the wind turbine as an autonomous unit with its own control system.

We have designed the whole system and have selected the subcontractors. The orders for the main equipment are yet shipped and the delivery times are scheduled by next May-June. We have also started phase II (construction and testing of the system on test bed).

Figure 4 : Wind turbine PEUI-10/2

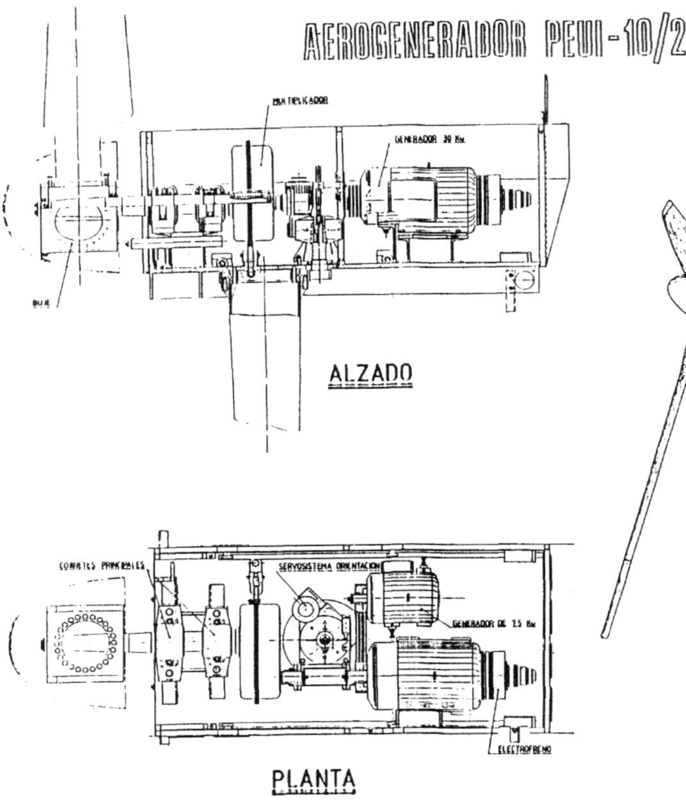

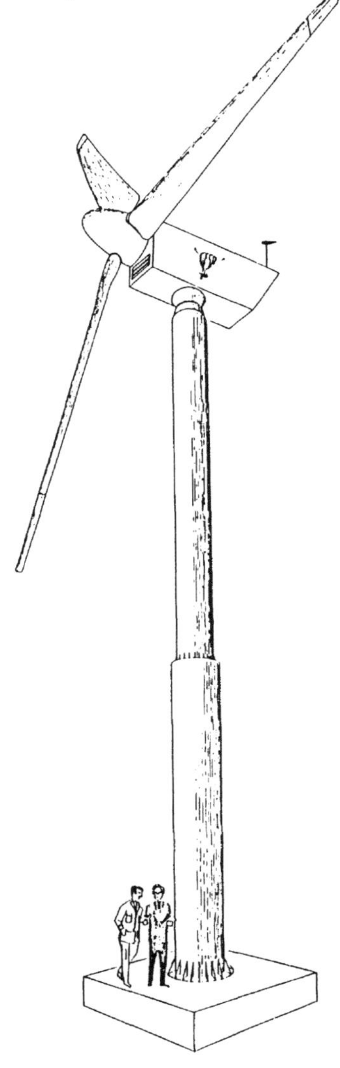

CARACTERISTICAS TECNICAS

ROTOR

Número de palas	3
Angulo de las palas	fijo (0°)
Material de las palas	fibra de vidrio
Eje de rotación	horizontal
Diámetro	10.4 m.
Velocidad nominal de rotación	70-75 rpm.
Sentido de rotación	agujas del reloj
Situación	barlovento
Freno	aerodinámico
Superficie barrida	$\simeq$ 85 m²

GONDOLA

GENERADORES	
Tipo	inducción trifásico
Potencia nominal	7,5/30 Kw
Potencia maxima	33 Kw.
Velocidad sincronismo	1500 rpm.
Factor de potencia (pot. nominal)	0.89
Rendimiento (pot. nominal)	87%
Protección	IP-54
Tensión	380 V (50Hz)
MULTIPLICADOR	
Tipo	ejes paralelos
Relación	1:20.3
ORIENTACION	
Tipo	servosistema eléctrico dirigido por veleta.
TORRE	
Tipo	tubular
Altura	12 m.
FRENO	
Tipo	electromagnético por reacción de muelles.

Figure 5 : Wind turbine PEUI-10/2

Project Nr 405/86 ES

AUTONOMOUS WIND-DIESEL PUMP SYSTEM

Pep Prats
Ecotècnia S. Coop.

Summary

The project aims to install an autonomous water pumping system for irrigation in rural areas. The contents of this paper is a description of the system and the present state of the project, on early developement.

1.INTRODUCTION

The system is based on the aplication of wind energy to water pumping, either from subterranean sources or from irrigation channels to higher altitude areas. The wind energy converter (WEC) and the pump are, most of the time, connected directly, that is, the transmission of generated power occurs without any intermediate conversion.

Up to now, the fixed pitch machine with stall regulation has proved the most reliable WEC at a minimum cost. This is a very important consideration in remote systems.

For this reason, the proposed system is based on this principe, that is, the direct connection of the WEC to the pump, so that there is no mechanism in the WEC to regulate the rotational speed. This is done by charging the generator from a certain number of revs on.

A support system driven by a diesel generator must be provided, so that the WEC can operate conjunctly as an energy-saving system, and, if necessary, it can run a supplementary pump together with the WEC when this is working with normal load at constant frequency, enabling an increase of the equipment's efficiency and profitability. The changeover between the two forms of operation is done according to the requirements and level of the storage tank, load demand and wind conditions.

2 DESCRIPTION OF THE VARIOUS COMPONENTS

Ecotècnia 12/25 W.E.C.

GENERAL DESCRIPTION: Horizontal axis, fixed pitch, propeller type, up-wind.

Rotor.

number of blades	3
diameter, m	12
speed, r.p.m.	65
rotation direction	clockwise
location, rel. to tower	upwind
type of hub	rigid
conning angle	0
tilt angle	7

Blade.	
total length,m	5.75
material	fiberglass reinforced polyester
weight, kg.	70
airfoil section	NACA 44 series
twist	30
tip chord, m	0.35
root chord, m	0.85
tip speed, m/s.	41
solidity	0.086
Tower.	
type	tubular steel tripod
tower material	steel
height, m	14
weight, Kg.	800
foundation	three concrete footings
Transmission.	
type	two stages
ratio	1: 23.6
input speed, r.p.m.	65
output speed, r.p.m.	1534
Main generator.	
type	syncron
rating, Kw	25
power factor	0.9
voltage, V.	220/380 Ac three phases
speed, r.p.m.	1500
Orientation drive.	
type	passive yaw
yaw drive	auxiliary propeller and mechanical transmission
Control system.	
rotor speed control	stall-regulated
overspeed control	fail safe hydraulic disc brake on main shaft
overspeed control	spoilers with manual reset
Performance.	
maximum power, Kw	25
wind speed at centerline of hub:	
start-up, m/s	4.5
cut-in, m/s.	4
max. prod, m/s	14
cut-out, m/s.	25
maximum design, m/s.	60

Annual power output (estimated).
6.7 m/s at 14 m. 69000 Kwh

Weight.

above tower, kg.	1500
total, kg.	2400

DIESEL GENERATOR

Mechanical capacity	26 h.p.
Electrical capacity	16 Kw.
Fuel consumption	0,2 kg Diesel/mech h.p.
Rotation speed	3000 r.p.m.
Cylinder capacity	1490 cm|3†"
Generator	Synchronous, 3 phase
Voltage	220/380 V
Optional extras	Automatic start

Pumps

Type	entrifugal, submerged
Electrical power	15 Kw

3. STATE OF THE PROJECT

At this moment the project is in the first phase, the site selection. Because the wind characteristics of the first site examined were not good enough we decided to look for another site.

That has caused a delay of about three months from original planning.

The first site has a high monthly variation of the average wind speed. Figure 4 shows the monthly average speed for the site, and the expected energy production. The wind energy contribution during the summer period has been considered too small.

This circumstance would cause a higher fuel consumption than desired.

We have checked other possible sites with better speed distribution, althrough they should maintain other site requierements: the installation must be easily reproduced and must be similar to other future applications.

After several site analysis, we have selected a site in Aragon, in a zone with better and higher wind speed averages than the first location. There is a farmer interested in using the installation.

The situation of the site is shown in figures 5 and 6. The average monthly wind speed and the wind direction distribution are shown in figures 7 and 8. The expected monthly energy production is shown in figure 9, together with the expected for the first site.

Figure 10 shows the actual bar chart for the project.

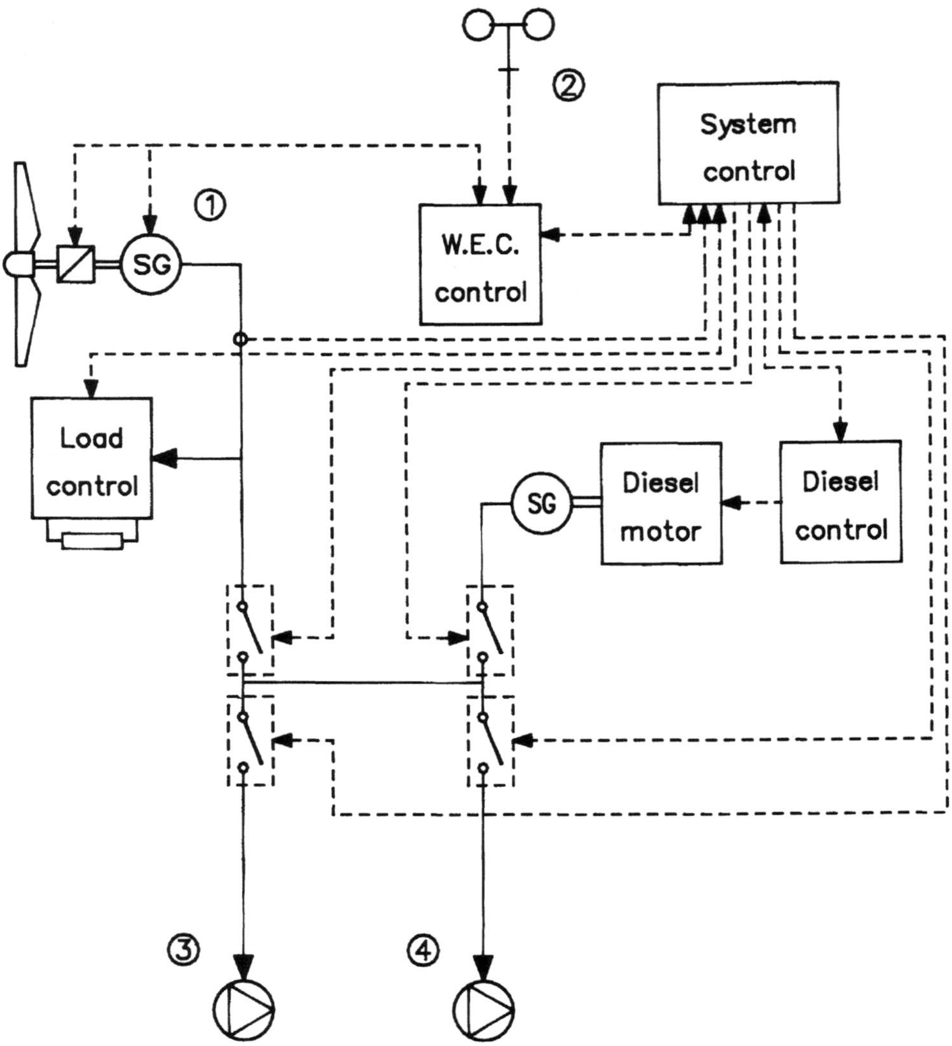

① WEC

② Anemometer

③ Pump 1

④ Pump 2

Fig. 1 — Functional diagram

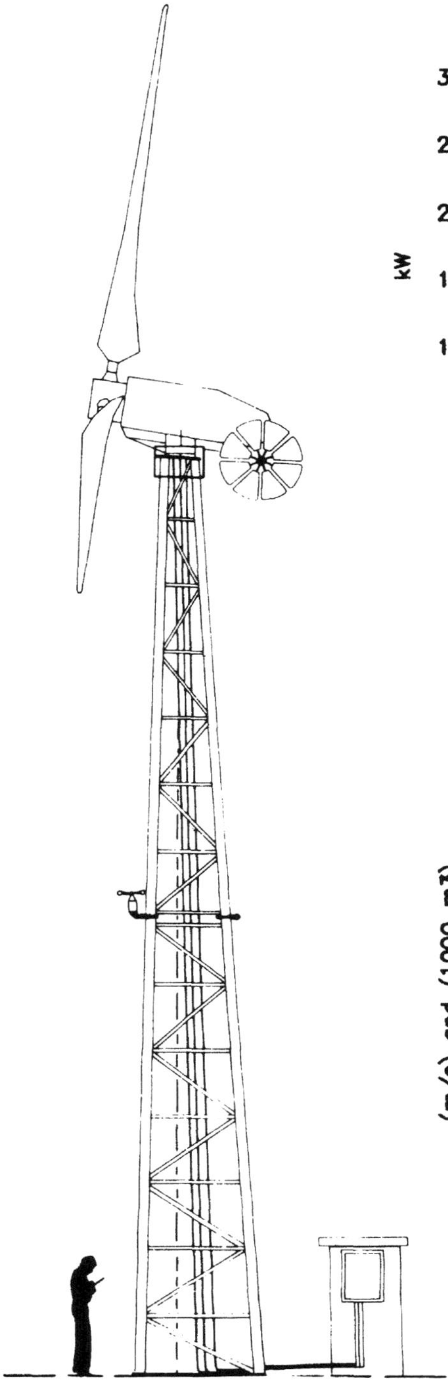

Fig.2. Wind energy converter overview

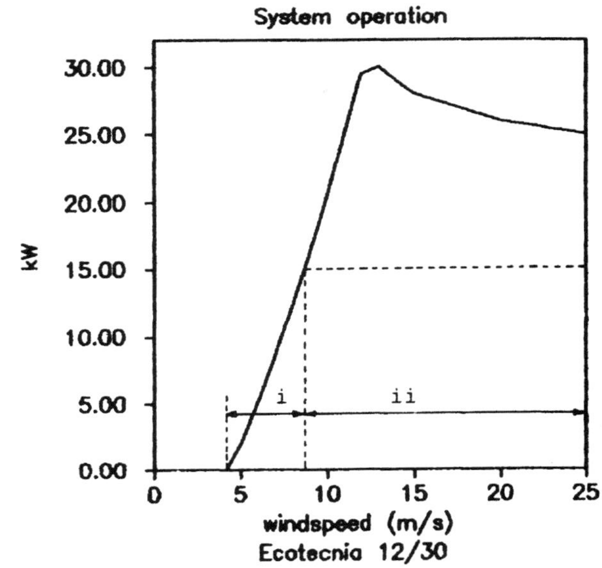

Fig. 3. WEC power curve and operation modes.
i - variable speed
ii- Constant speed

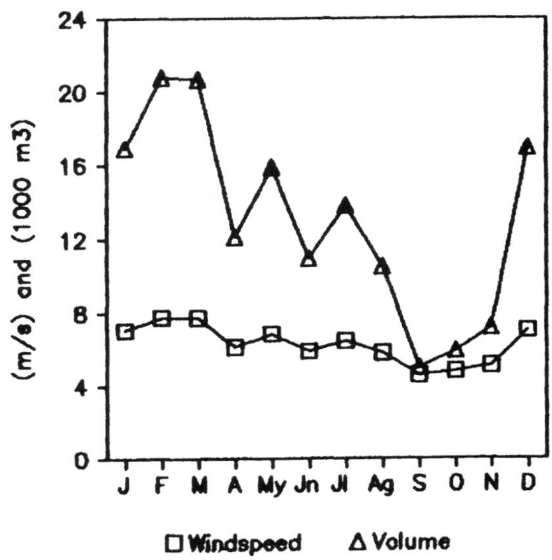

Fig. 4. Monthly windspeed and pumped volume in the first site in La Mancha

AUTONOM. WINDPUMP SYSTEM

Site location near Zaragoza

Data: 15.4.88

Map 1

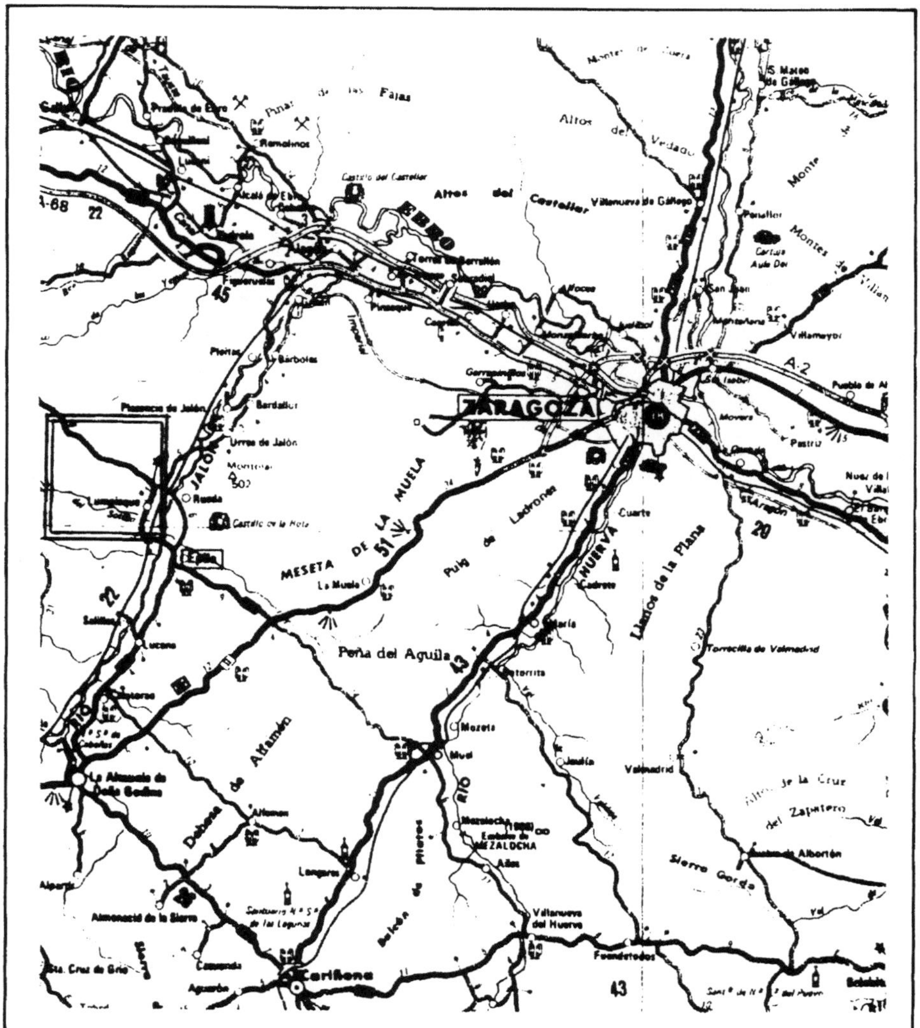

Fig. 5

AUTONOM. WINDPUMP SYSTEM	
Detail of site location	Data: 15.4.88
	Map 2

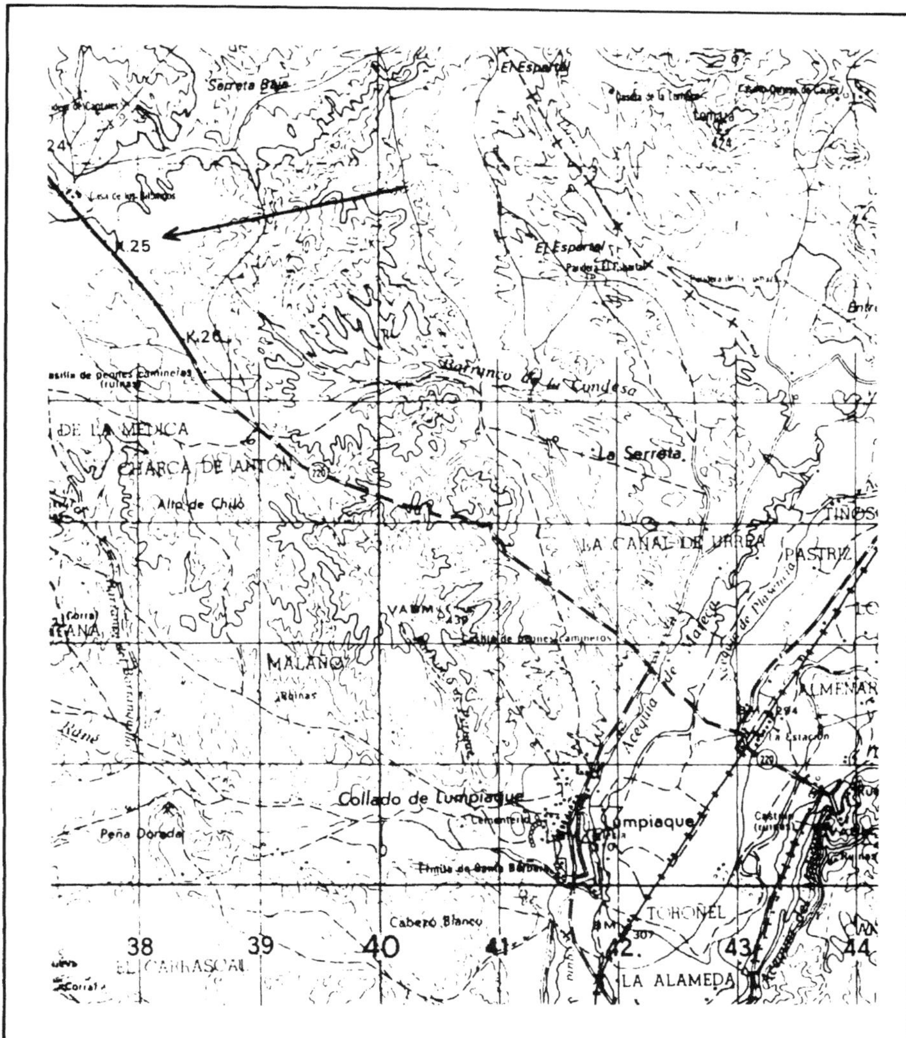

Fig. 6

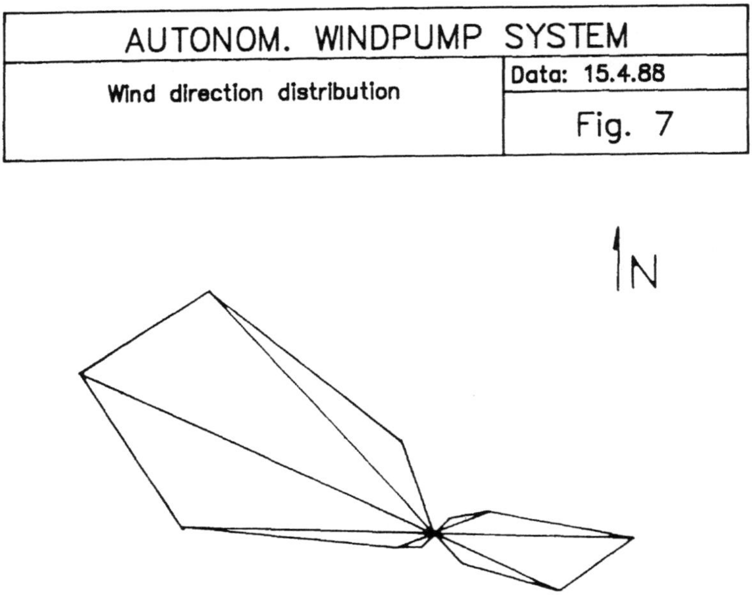

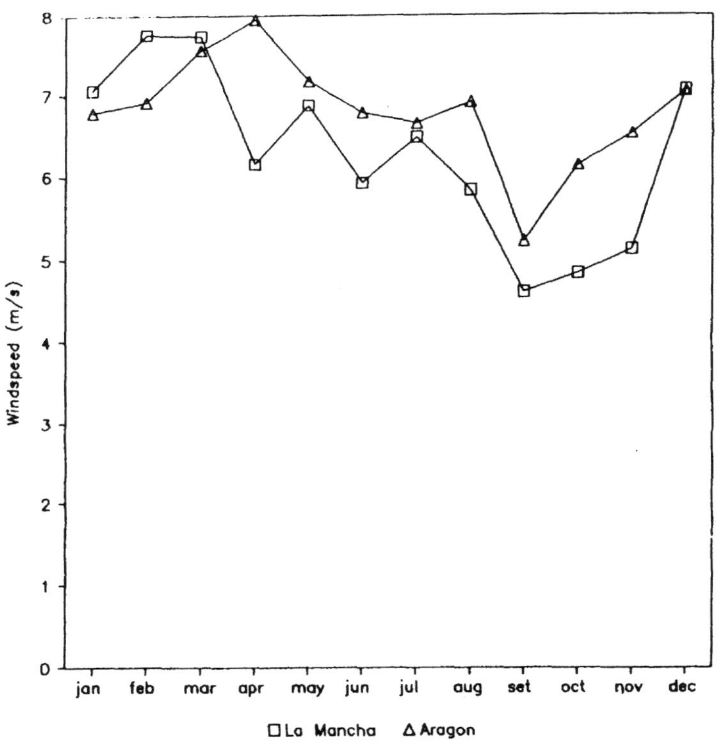

Figure 8 : Monthly average windspeed in both sites

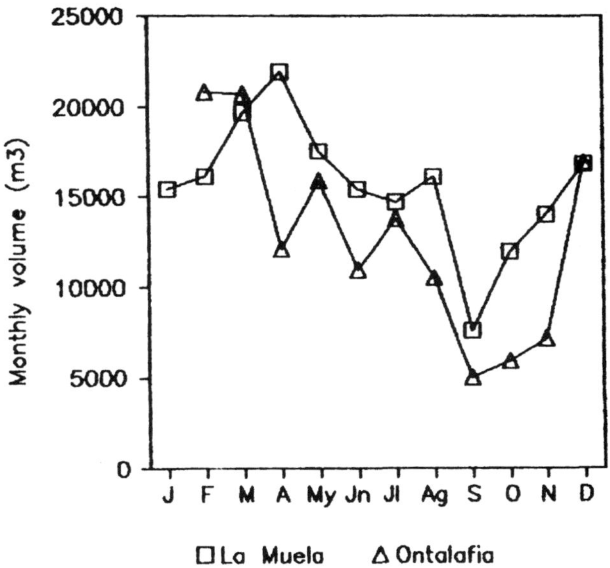

Fig. 9. Monthly pumped volume in both sites

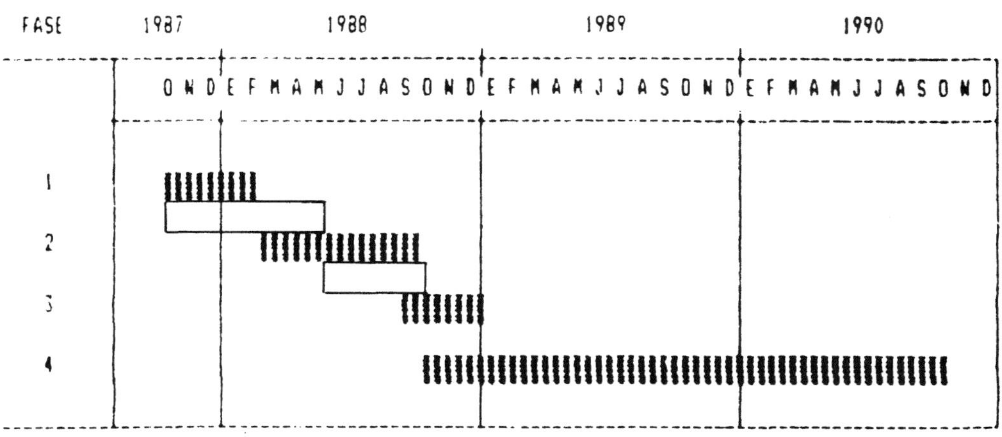

Fig. 10. Actual bar chart for the project
1. Site location
2. Installation
3. Start up
4. Testing

Project Nr 512/85 UK

FOULA WIND-PUMP-HYDRO SYSTEM
THE DEVELOPMENT OF A CONTROL STRATEGY

W.M.SOMERVILLE; W.GRYLLS; G.D.NICHOLSON;
G.R.WATSON; M.D.JEPSON
Windharvester Ltd, Hexham, U.K. NE46 4SA

Summary

Control of the three generators and three pumps is exercised from the supervisory controller located at the diesel generator house. Communication between the three sites is provided by a telephone cable with five twisted pairs which provide for safety, control, communication and datalogging. Remote control of the wind and hydro generators can be accomplished using an enable signal from the supervisory controller (in manual or auto modes). Starting of any generator would of course be inhibited should a fault be present on that generator. All civil work on the project, including buildings, dam works and penstock is complete. The distribution transformers, HV and LV cable network is installed. Control cabinets and wind turbine have been built and tested and are presently being shipped to the island.

1. INTRODUCTION

The Island of Foula, part of the Shetland Islands group, lies in the Atlantic some 50 kilometers due West of the town of Lerwick. The small population, currently 48, and the isolation of the Island by some 20 kilometers of open sea mitigates against any future possibility of an electricity supply by undersea cable from the Shetland mainland.

The scheme to be commissioned Summer 1988 relies on energy provided by a wind turbine generator, equipped for stand-alone operation, as a primary source of power augmented by a small natural contribution from a micro-hydro scheme, capable of independent operation or in parallel with the wind turbine, and which can also act as a pumped storage facility to absorb surplus wind energy against periods of calm. A back-up supply is provided by a diesel generator to ensure the availability of essential services in the event that wind and hydro resources are exhausted or inoperative.

The locations of the generating plant and the Island dwellings are widely dispersed along the eastern side of the

Island and the provision of a community electricity service involves a lengthy distribution system and necessitates the use of high voltage cables and transformers.

Overall system management in automatic operation will be provided by an industrial grade micro-processor system. This controller will maintain watch on a number of sensors monitoring the status of the system and order the management of the system according to one of several preprogrammed strategies as selected. The operating strategy will be constructed to evaluate the benefits from operating the system in different ways to achieve required objectives for service, such as for example, minimum use of diesel fuel, or conversely, the maximum hours of service.

The standard control computer is equipped with both analogue and digital input and output interface units to give adequate isolation and protection to ensure reliable operation. The central processor and the input and output modules are arranged on a back plane assembly so that items may be readily changed, or replaced, should the need arise.

Communications between the central control unit and the sensors at the remote generating sites will be provided by a signal multiplexing system. Switch systems can provide up to 64 communication channels along a single twisted pair of telephone line at distances up to 10 kilometers, simultaneously handling both analogue and digital signals.

2. GENERATING PLANT

The primary source of energy for the scheme will be provided by a wind turbine with a nominal rating of 60kW equipped for stand-alone operation. The wind turbine rotor will be of the fixed pitch "stall regulated" type and will be equipped with a synchronous alternator of the brushless self-excited type with a nominal rating of 97kVA. The alternator will generate 3 phase 4 wire output at 415/240 Volts controlled by its own automatic voltage regulator. The nominal frequency of the supply generated will be 50 hertz and the necessary control of frequency will be achieved by an automatic load management scheme, sensitive to minor changes in the generated frequency, which will automatically apply load to appliances or the wind turbine local dump loads in order to continuously match the applied load to the power output of the turbine.

The secondary source of energy for the scheme will be provided by a micro-hydro generator with a nominal rating at maximum output of 23.5kW. Hydro power will be drawn from a natural reservoir with an initial capacity of approx 1 million gallons at a head of 110 metres generating power through an impulse turbine controlled by a variable spear valve. The impulse turbine will be directly coupled to the 4 pole brushless self-excited alternator and, in addition, coupled to an extension of the alternator shaft an 11kW induction generator will allow rapid coupling of the water turbine to the system when the wind turbine is unable to support the system load. The alternator will have a nominal rating of 48.6kVA and be provided with its own automatic voltage regulator with quadrature droop control for parallel operation in synchronism with the wind turbine as well as offering the facility to run the network unaided.

The water turbine, although equipped with a spear valve, will operate ungoverned, the necessary frequency stability for the system being provided by an automatic load management scheme functionally identical to that provided for the wind turbine with a local dump load. This arrangement ensures that the governing system on the two prime movers is inherently compatible over the entire range of permissible settings thereby avoiding the complexities of providing a mechanical governor with suitably adjustable characteristics for this function. Although ungoverned the load control system controls the set position of the spear valve to match the essential service demand, and, subject to the control strategy selected and the available water reserve, will operate at a margin which allows a variable secondary level of power for heating. Thus no hydro power will be lost and the saving in reduced complexity by eliminating the governor and its special maintenance requirements is worthwhile.

These two generating schemes may operate independently to provide the Island power, or may operate in parallel as determined by the prevailing conditions and the control strategy selected for system operation within the following limits.

Wind turbine rated power at 14m/s wind speed	60kW
Wind turbine dump load rating	100kW
Water turbine rated power at 110m head	23.5kW
Water turbine dump load rating	25kW
Generated voltage 3 phase balanced load, power factors 1.0 to 0.8 lagging	415/240V +/-2%
Nominal frequency of supply	50Hz
Frequency range for distributed heating load control	50.1-51.5Hz
Frequency range for automatic dump load application	52.0-53.5Hz
Overfrequency shutdown initiated	53.5Hz
Underfrequency disconnect of service power	47.6Hz
Underfrequency voltage reduction below	47.5Hz
Underfrequency system shutdown	45 Hz

In the event that wind or water power are unavailable or disabled a standby generator is provided driven by a 1500rpm diesel engine with a rated electrical output of 23.4kW. This generator will be subject to time clock control and only operable within certain agreed guaranteed supply hours.

3. SYSTEM SAFETY

System safety is provided for by the use of a 24V d.c. signal emanating from the safety battery at the diesel house. This circuit must be complete for any of the generators to operate on or off the network, (except to operate off the network under manual local control). The safety circuit may be broken by emergency stop buttons at each generating station. The earth leakage sensing relay will also break this circuit under a fault. In case of this circuit being broken, each generator must be locally, manually reset. Thus de-energisation of the safety circuit immediately disconnects all generators from the network and initiates system and generator shutdown.

One twisted pair in the control cable is dedicated to voice communication by telephone, between jack points provided at the local control panels and where appropriate to aid the initial set up commissioning and testing.

4. OPERATING MODES SINGLE GENERATOR OPERATION

4.1 Wind turbine

With the system dead, provided there are no local faults and the safety circuit is energised, the wind turbine will run up when there is sufficient wind, until loaded against its dump load.

The power produced will be monitored for a short interval until it has sustained a pre-set level, adequate to energise the system, for a pre-set time. At this stage provided there is an enable signal from the diesel shed, (and as an additional safety lockout, no voltage on the distribution system), the local controls will close the export contactor to energise the distribution network.

The energy produced by the wind turbine will be exported first to provide for lighting on demand, and then the second priority to provide domestic power on demand subject to a maintained frequency level with the residual power being available to provide for automatically controlled distributed heating loads. Any power surplus to the available load will be dissipated automatically in the local dump load at the wind turbine. If there is insufficient energy, the frequency of the output will drop, shedding first the heating load, then the essential service power and ultimately the lights will be disconnected when the frequency falls to 45 Hz. Sensors at the wind turbine will report power produced and power being discarded to dump to the supervisory controller.

4.2 Water turbine

On receiving the command to start up the water turbine control system will check that there is water present at a safe level in the reservoir and that the penstock is full before opening the spear valve to a first pre-set starting position. This valve setting will be sufficient to ensure a reasonably prompt start-up and acceleration to required running speed and when this is obtained the machine will run into its local dump load and if necessary a further opening of the spear valve will take place to raise the power generated to a pre-set level adequate to energise and sustain the network. At which point, providing the local controller finds that the network is un-energised, the export contactor will close.

Once the water turbine is powering the system, the power controlling spear valve may be opened further subject to the available water level in the reservoir to increase the output to provide domestic heating in addition to basic services, but avoiding spillage of power to dump. As the water level in the reservoir falls towards the minimum level, the power produced by the water turbine will be throttled back by the spear valve progressively, until only just sustaining the basic services as detected by frequency sensitive relays.

The water turbine will report its condition, power level and water level, to the central controller.

4.3 Diesel generator

Operation of the diesel generator in the absence of wind or water power will be subject to time-clock control providing two set periods of generation each day. When enabled,the diesel generator will start automatically during the permitted running hours at any time when the service supply is lost from the distribution system. Up to 3 start attempts will be permitted and if all 3 fail the machine will lock out. When the diesel generator is up to speed it will be allowed to run for a period of 60 seconds to warm up, and providing the supply has not appeared on the network will automatically energise its export contactor to provide power. Should the power supply be restored during the diesel start sequence, the diesel will be allowed to run until thoroughly warm before being shut down. The diesel generator will report its status and condition to the central monitoring and the supervisory control unit.

5 OPERATING MODES COMBINED GENERATOR OPERATION

5.1 Water turbine closing onto wind-powered system

There are two conditions when the water turbine will be called upon to co-generate with the wind turbine. One, when the wind turbine is low in output and close to dropping the essential services. Two, when the reservoir is full and about to overflow and the wind turbine is not providing enough power to fully power the system as determined by the export power level or by the absence of wind turbine dump load.

On receiving the command to start from the central controller the water turbine start-up sequence begins as previously, but in this case the presence of power on the local distribution network automatically engages a starting sequence to lock in the induction generator when synchronous speed is reached. This ensures that a power contribution from the water turbine is immediately available without any delay necessary to achieve synchronisation,important when the wind turbine is failing to deliver sufficient power for the network. Once the induction generator has engaged, the relative slip between the alternator on the water turbine and the network is reduced to about 1 - 1.5% fast. Closing the induction generator to the network initiates a phase matching sequence detected by a relay which subsequently closes the contactor to synchronise the water turbine alternator to the network. When this sequence has completed the induction generator is automatically disconnected and the spear valve control will move in order to maintain the service frequency and output power at the desired level subject of course to the available reserves of hydro power in the reservoir.

The two machines will continue to operate in synchronism either to increase the net power to the system when there is adequate water power or to maintain basic services when water resources are limited.

If the input from the wind generator should continue to fall its power output will be monitored and should it become negative for a set interval, the wind turbine will be automatically disconnected and allowed to run to its local dump or to standstill. In this event the water turbine will take over the network load and continue to operate subject to the water conservation strategy adopted.

If during parallel operation the wind turbine power should increase to the point where power is being wasted to dump, the water turbine output will be reduced if necessary to zero in order to conserve storage. Further reduction in the water flow will require a net input from the network to the water turbine alternator to maintain it spinning as a synchronous condenser, and if this condition is maintained for a preset interval the machine will automatically disconnect and go into a standby condition.

5.2 Wind turbine closing onto a water powered system

If the system is powered from the water turbine during a period of calm and the return of wind permits the start of the wind turbine, the wind turbine will run up and automatically load against its own dump load. It can not connect to the network which is already energised from the water turbine. The supervisory controller will monitor the output from the wind turbine until it has maintained a preset level for a preset time.

At this stage the supervisory controller will change modes of operation to wind turbine and hydro in parallel. It will accomplish this in the following manner. First under a command from the supervisory controller the hydro will open its synchronous contactor. As soon as the network is dead the wind turbine will close its contactor. When the network voltage reappears, due to the wind turbine, the hydro local control will connect its induction generator to the network. (In the short interval when the hydro was off the network it would have been running into its local dump load.) As soon as synchronism is achieved the hydro local control will connect its synchronous generator and disconnect the induction machine.

The wind and hydro turbines are now running in parallel on the grid. The supervisory controller will be monitoring the power outputs and will command a hydro turbine power output level according to water levels, guarantee periods and the power provided. If wind power output is sufficient on its own the supervisory controller will command the shutdown of the hydro turbine.

It is not intended at the present time to provide for operation of the diesel generator in parallel with either the wind turbine or the water turbine, since at this point in time a major objective during the initial period of operation is to establish as quickly as possible techniques which will avoid or at least minimise the amount of diesel fuel used for electrical power.

5.3 Pumping to Store

System priorities remain first to provide a service and secondly to conserve as much hydro resource as is practicable consistent with a reasonable level of service.

In order to minimise the amount of energy lost to dump during the operation of the system 3 pumps have been provided to lift water from the West Burn to the Ovrafandal Reservoir. The first pump is a submersible suction pump with strainer, which raises water some 5 metres from the West Burn to a holding tank at the hydro station, with a power rating of 2.25kW. Two further multi-stage centrifugal high pressure pumps each rated 5.5kW are provided at the hydro station to raise water from the holding tank through non return valves via the penstock to the dam.

Pumping will be initiated when the central control system records the fact that a significant level of power is being dissipated in dump by the wind turbine. The sequence will provide for the starting of the initial feed pump followed when the holding tank is full by the initiation of the first of the high lift pumps. The operating conditions and the amount of power to dump will continue to be monitored and if there is still power being wasted the second high lift pump will also be energised. Once a high lift pump has been started it will be allowed to continue to operate subject to the supervisory system confirming that there is adequate power (i.e. above a set power level from the wind turbine) regardless of whether power is being dissipated in dump or not.

Pumping shall not be permitted if the reservoir level sensor indicates the reservoir is already full, or close to its full level. If during periods when the diesel generator is operating and the delivered power is very low it is permissible to pump water to improve the load factor on the diesel generator providing the reservoir levels are low.

6 TRANSMISSION AND DISTRIBUTION

The shortest practical route between the wind turbine generator and the most remote property to be supplied exceeds 6,000 metres, and the total length of cable required for generator interconnection and distribution to the properties designated as requiring a supply totals some 9,400 metres. To keep volt drop, cable cost and the weight of cable to be handled to acceptable levels a high voltage distribution system is required.

Distribution/Interconnection will be provided by a 3,300 volt 3 phase cable some 5,100 metres in length with 9 step down transformers to provide the normal 415/240 Volt service.

The scheme provides for five 3 phase supplies to workshops at North Biggins, Niggards, Groups, the Harbour, and also to the public water supply pump house. All other supplies will be fed single phase with phase selection as appropriate to keep the system in near balance. Provision will be made to provide a three phase supply to the new School and Community building adjacent to the diesel generator building from the system.

6.1 Three phase terminals

The feeder cable will terminate in a 4 way splitter box carrying 3 sealed fuses, one for each phase, and a link in the neutral. Connection will then be made directly to a three phase or three single phase meters as available. In addition, a single phase tapping is taken from the output to a load management unit box which will permit a direct feed to the lighting distribution fuse board of the customer's supply and will in addition provide a frequency controlled outlet of a maximum capacity of 2kW for local heating with a second control output which powers the contactor enabling the three phase supply.

The set frequency of the heating circuit will be dictated by the management committee for the system and will operate automatically subject to the consumer's desire to have the heating circuit enabled by means of a manually controlled switch. The second frequency control circuit will normally be set at 47.6 Hz and will thereby permit operation of 3 phase appliances at all times when there is sufficient wind or water energy to provide the necessary power. When this circuit is energised the 3 phase output contactor may be closed by the operation of a manual push-button and held in by a self-sealing contact on the contactor. If at any time the applied load causes the frequency to dip below the set point the contactor will be de-energised and because of the broken self-sealing contact will not re-close until re-set manually.

The provision of this combined control scheme ensures that power tools will be disconnected from the supply before the system is disconnected by the overall under-frequency trip and thereby the lighting should remain on for a sensible time after the power has been removed from the power tools.

6.2 Single Phase Terminals

For all domestic consumers a single phase supply is provided by underground cable terminating in a line fuse and neutral link which is directly connected to the primary energy meter which records total kW hours consumed. A second energy meter only records power consumed when the frequency is low i.e. less than 50.1 Hz signifying that the diesel is in use. An extension of the supply system passes through a third kWh meter to the load management unit which senses the frequency of supply and provides a driving signal to control the second kWh meter and the power relay and provides for up to 4 heating circuits automatically controlling heating loads, such as water heating and space heaters in response to pre-set spot frequencies in order to utilise excess energy from the wind turbine.

The priority established between the first circuit for lighting and the second circuit for power on these domestic units will ensure that the last resource of the wind turbine will be available for lighting purposes. The power contactor controlling the power circuit is subject to a time interval so that should there be a momentary disconnection due to insufficiency of supply against demand, the power circuit will be disconnected and not reconnected until a preset time has elapsed (2-5 minutes) even if sufficiency of power is re-established almost immediately.

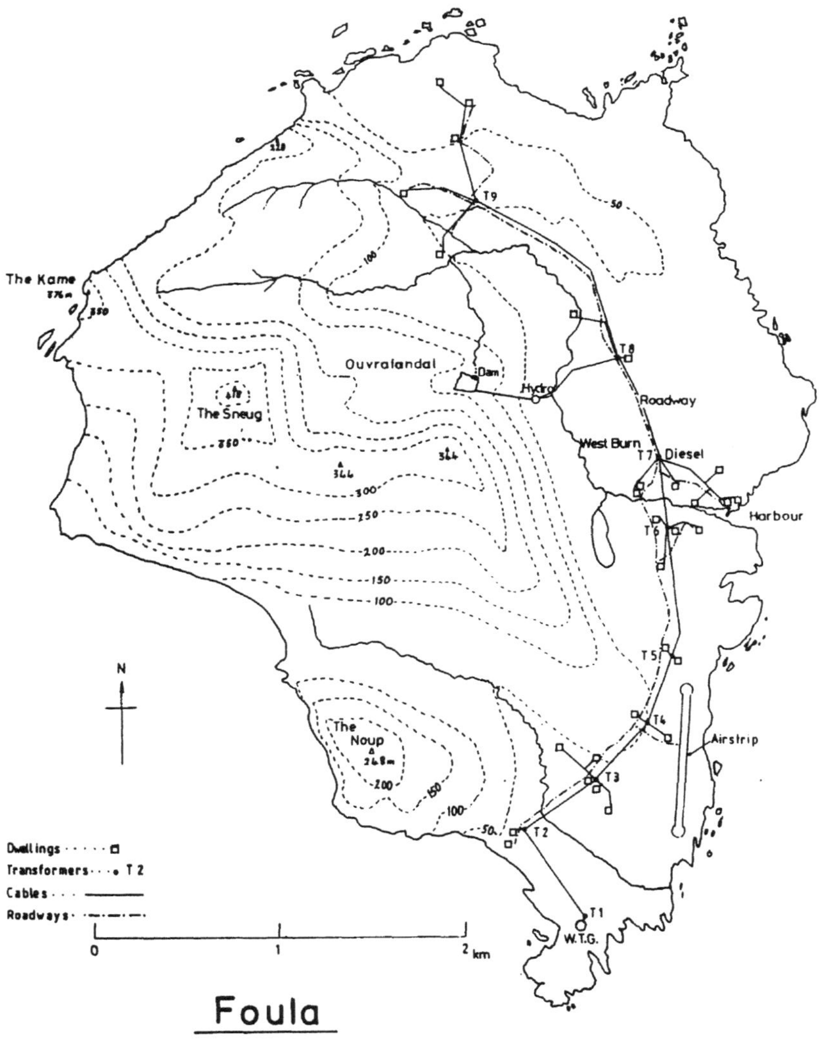

Figure 1 : FOULA PHYSICAL LAYOUT

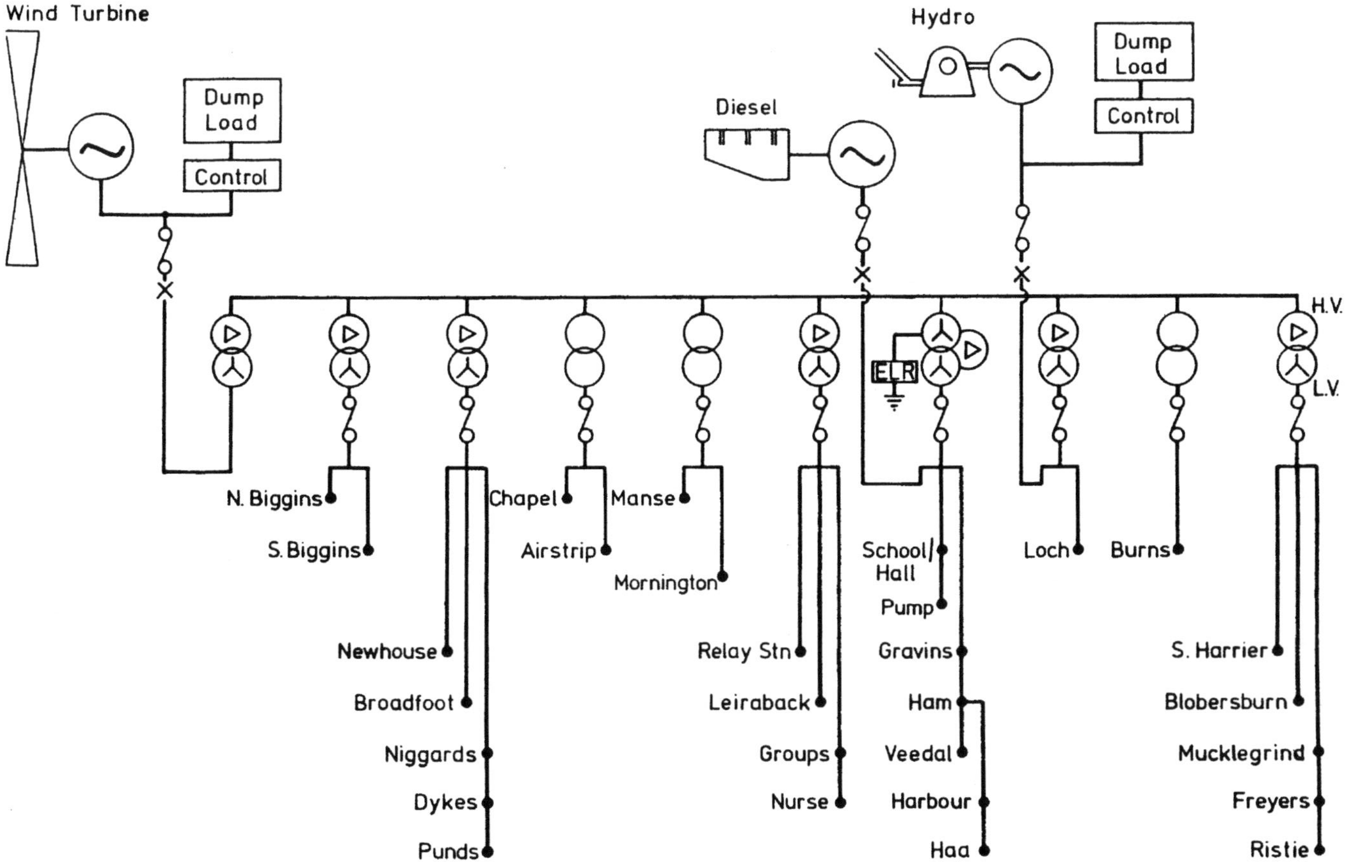

Figure 2 : ELECTRICAL SYSTEM FOR FOULA

LEGEND

C = Four pole power contactor
M = Energy meters (kWh)
MCB = Miniature circuit breaker
PA = Power available neon
POC = Power oscillator card
PR = Power reset pushbutton
RCCB = Residual current circuit breaker
TC 1 = Timing card no:-1
TR 1 = Triac no:-1
UFC = Under frequency card
UFR = Under frequency relay

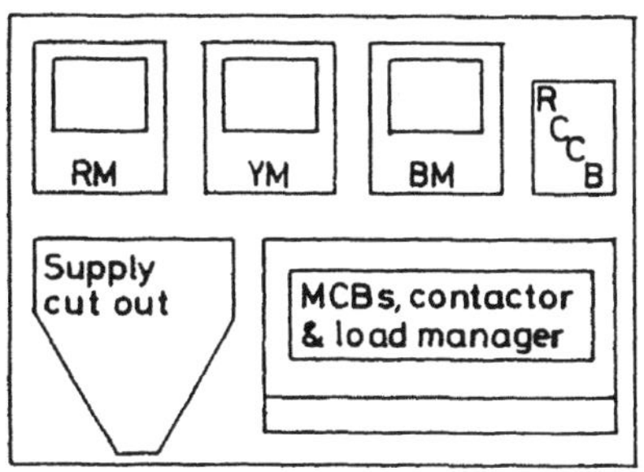

Meter board layout

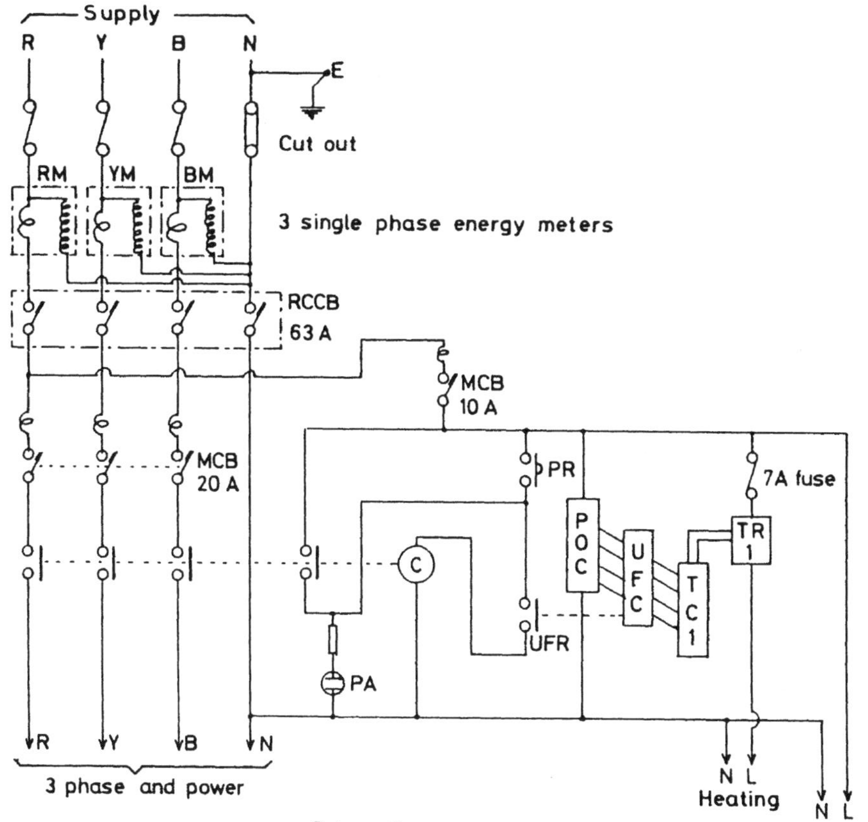

Schematic

Figure 3 : THREE PHASE TERMINAL SCHEMATIC

LEGEND

C - Contactor 20A. power
M = Meter kWh
MCB = Miniature circuit breaker
POC = Power oscillator card
RCCB = Residual current circuit breaker
T = Time delay relay
TC 1 = Timing card (programmed)
TR = Tariff relay
TRC = Tariff relay card
TR 1 = Triac no.1 (solid state relay)
UFC = Under frequency card
UFR = Under frequency relay

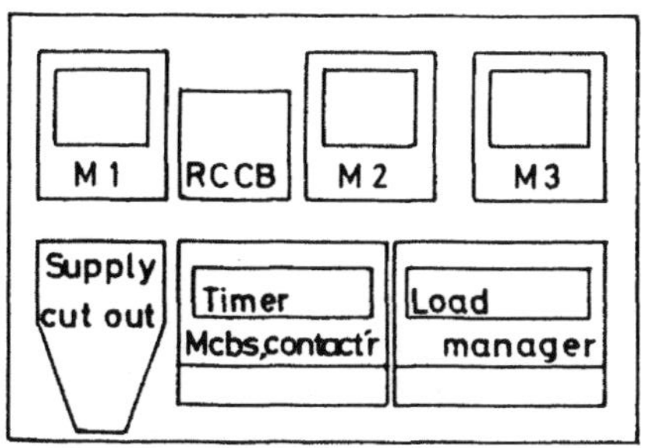

Meter board layout

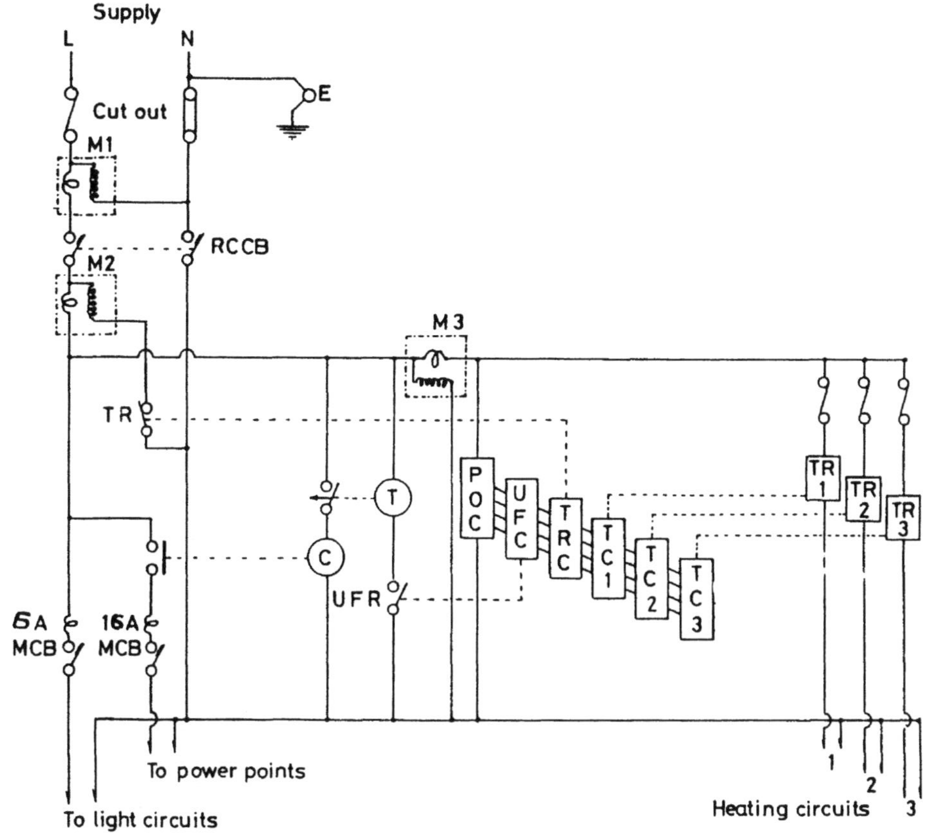

Schematic

Figure 4 : SINGLE PHASE TERMINAL SCHEMATIC

WIND DIESEL PROJECT IN CHINA

Svend-Erik Andreasen
Head of Energy Technology Department
COWIconsult, Consulting Engineers and Planners A/S
45 Teknikerbyen, DK-2830 Virum, Denmark
Telephone 45 2 85 73 11, telex 37280 cowi dk.

Summary

COWIconsult, Consulting Engineers and Planners, with RISØ (National Laboratory) as sub-consultants, has joined forces with BONUS (Danish wind turbine manufacturer) in a wind diesel project in China. The main tasks of the Danish team are planning, delivery, commissioning, supervision and monitoring of a wind diesel pilot plant and a testing centre. The project includes development of a joint Chinese/ Danish wind diesel system and training of Chinese engineers in wind technology during courses in China and Denmark.

1. INTRODUCTION

The project is supported by the EEC - DG8 on behalf of the State Science and Technology Commission of China and the project is just now in the initial planning phase.

The receiver of the project is the Zhejiang Provincial Science and Technology Commission, and the project is going to be installed on a small island called Xia Dachen Island (figure no. 1).

RISØ, National Laboratory, BONUS, windturbine manufacturer and COWICONSULT consulting engineers and planners, all from Denmark are joining forces in this project.

The main tasks of the team to be carried out in the future in collaboration with Chinese colleagues are:

A. Wind Farm and Control System
B. Wind Diesel Testing Centre
C. Development of Joint Wind Diesel System

Time limits are indicated on the activity time schedule (figure no. 2).

Re.: Task A, Wind Farm and Control System

This phase comprises design and installation of a small wind turbine farm and a wind diesel control system, including training in operation and maintenance. The farm will consist of wind turbines connected to an existing grid system on the island which is representative for small grid systems on other Chinese islands. The farm is considered as a pilot wind farm and experiences achieved in connection with the implementation and operation of this farm can be transferred to future similar wind turbine installations on Chinese islands and in the Pacific Region in general.

During this phase the total farm will be designed and erected based on detailed information concerning the site, wind conditions, power demand and the conditions of the existing power/grid system. At present a team of engineers is in China for fact-finding.

The proposed system consists of wind turbines within an installed capacity of approx. 150 kW, and equipment for control of the grid frequency and load conditions.

The following information about monthly wind speed average and expected average load curves has been received from China (figure nos. 3 and 4).

The system will be supplied with a central control computer provided with software for:

- automatic control of the wind turbines.
- automatic control and load management of the existing diesel generating sets.
- collecting and reporting of all main process data in the system.

The effect of the existing diesel power plant is approx. 800-900 kW.

The central control computer will be able to control operation of the proposed system as a wind diesel system and as a stand-alone system.

Re.: Task B, Testing Centre

The general layout of the wind diesel testing centre will mainly be determined by the specified test and measuring tasks to be carried out at the testing centre. Three main tasks for the testing centre have till now been identified.

They are:

1. Testing and Measurements

- Detailed testing of WEC's at two testing platforms
- Detailed measurements of the operational characteristics of the coming wind diesel system
- Long time monitoring of WEC's and diesels combined with measurements of meteorological parameters.

2. Establishing Wind/Diesel Know-How

- Computer analysis of Wind/Diesel systems in parallel to the measurements.
- Exchange of programs/experience with other Wind/Diesel test centres.

3. Consulting on Wind/Diesel System Design and Application to:

- Manufacturers
- Users
- Planners and other authorities.

The layout of the wind diesel testing centre shall ensure that different activities (measurements and analysis) can be carried out simultaneously. The flexibility of the system shall also allow that the system easily can be extended to include measurement on another wind diesel system if such a system is installed at the site.

Wind Turbine Testing

The basis for the testing of wind turbines will be the evaluation of the test site. The site will be modeled with a computer program which is able to describe the effect of the wind flow in hilly terrain. This program is a part of the European Wind Atlas Program.

Wind Diesel Testing

On basis of the experience gained at the wind diesel test facility in Denmark a measurement programme will be set up for the evaluation of the operational characteristics of the Dachen Islands wind diesel system. In particular the influence of alternative control strategies on the specific fuel consumption will be investigated. For such investigations the present system is well suited as the diesel power plant including the five individual diesel generator sets allows a continous notching of the diesel power to the actual wind power and load consumption.

The measurements of power quality (variations in frequency and voltage) should also be mentioned as one important part of a wind diesel test programme which will be included.

Load Simulator

The wind diesel system will include a dump load. For testing purposes this dump load will be coupled as a load simulator.

Re.: Task C, Development of Joint Wind Diesel System

This phase comprises an evaluation of European and Chinese WEC's and wind diesel control strategies.

Based on a market analysis for wind diesel systems carried out and through an analysis of Chinese manufacturing capabilities a strategy for commercial production shall be developed.

2. CHINESE MANUFACTURING CAPABILITY AND CAPACITY

When evaluating the Chinese manufacturing capability, several issues must be covered, e.g.

- Which components can be delivered from Chinese manufacturers with sufficient quality control.

- Which components of Chinese origin could be used if the quality could be improved.
- What is the depth of manufacturers for individual components. If one manufacturer should prove unable to deliver, can the supplier of the component be changed?
- Which components could be manufactured by Chinese firms now producing similar products.
- Where are agglommerations of firms producing the necessary components in relative proximity. How far are these from the future markets.
- Can relevant Chinese firms be induced to go into joint ventures for the production of wind turbines?

China is a huge country with an industrial sector as large as that of many European countries. China produces a wide range of industrial products including a variety of products of the kind necessary for or related or similar to the production of wind turbines. There are also Chinese manufacturers who have produced large numbers of small wind mills. These firms might be interested in taking over the production of larger units. Therefore, the difficulty of evaluation the Chinese manufacturing capability and capacity will be in the filtering and selection of cooperation partners according to the needs of the wind turbine production process, that is, under the assumption that there is interest in such a production in the firms.

In order to make an effective technology transfer all activities will be carried out by a Joint Wind Diesel Work Group.

Joint Wind Diesel Group Activities

As illustrated a Joint Wind Diesel Group (JWDG) is to be set up consisting of 3-4 Chinese engineers and 3-4 European experts (figure no. 5). The engineers and experts should cover know-how within WEC-technology, wind diesel system planning, equipment manufacturing, market and economic analysis.

The JWDG will follow the design and implementation of WEC's and control system and will be responsible for

- Planning of monitoring programme. Decision as to parameters to be monitored and data evaluation principles.
- Data analysis and evaluation.
- Evaluation of WEC project
  - Technical and operational performance
  - Cost determination
  - Power production and consumption, power quality
  - Reliability
  - Performance of Chinese WEC's vs. European WEC's.
- Recommendations on and implementation of system concept improvements
- Development of Joint Wind Diesel concept for commercial production
- Detailed planning of all training programmes.

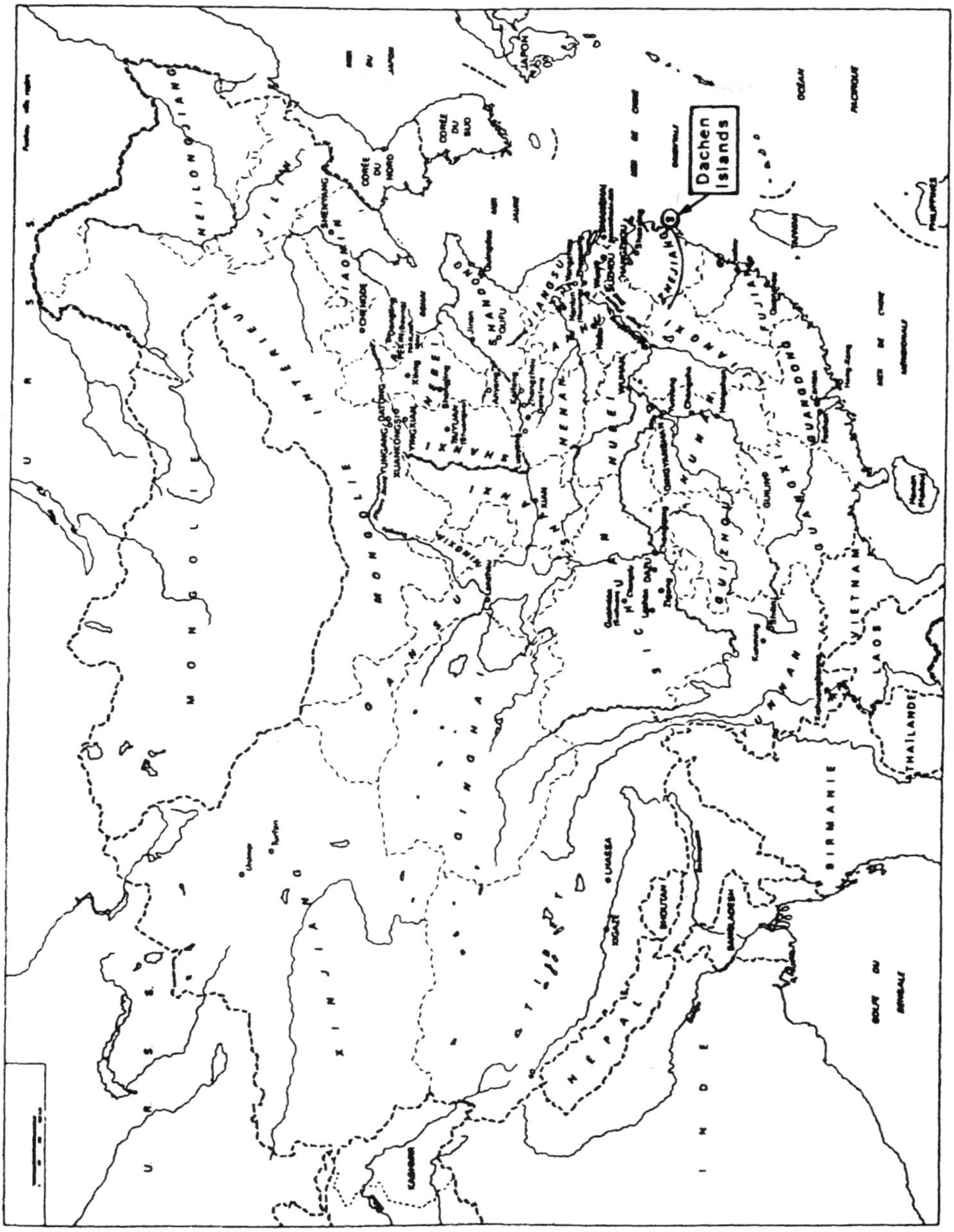

Figure 1

Activity Schedule

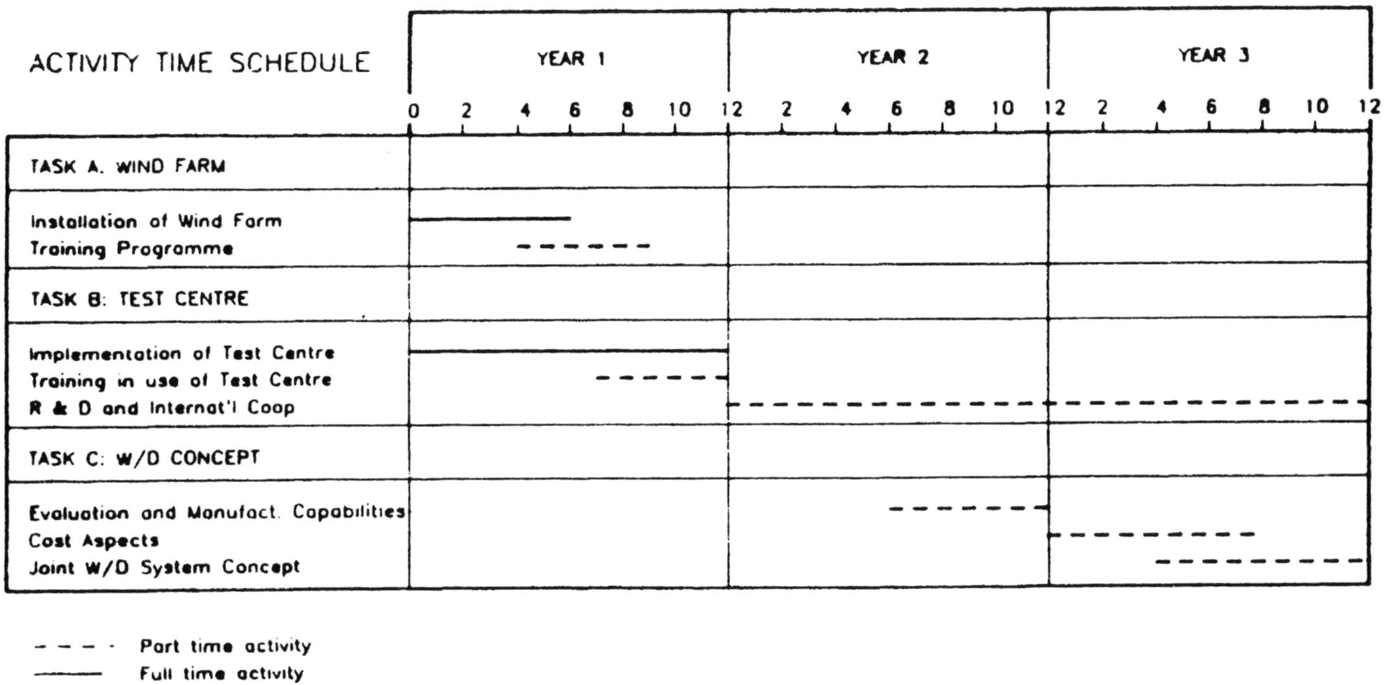

Figure 2

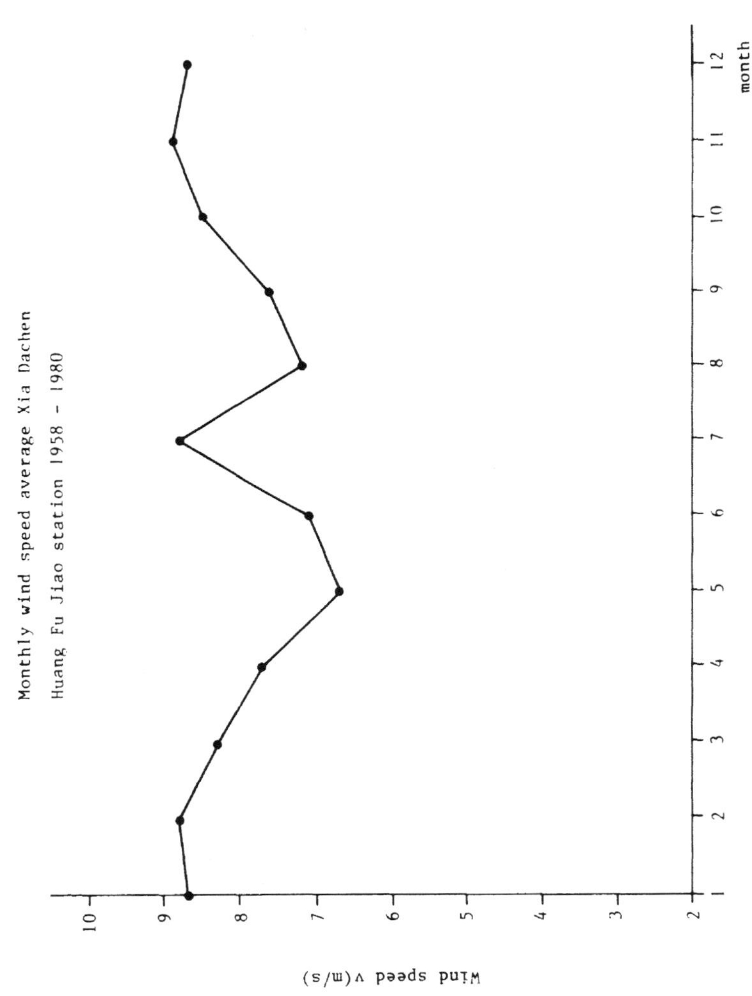

Figure 3

Monghly average wind speeds

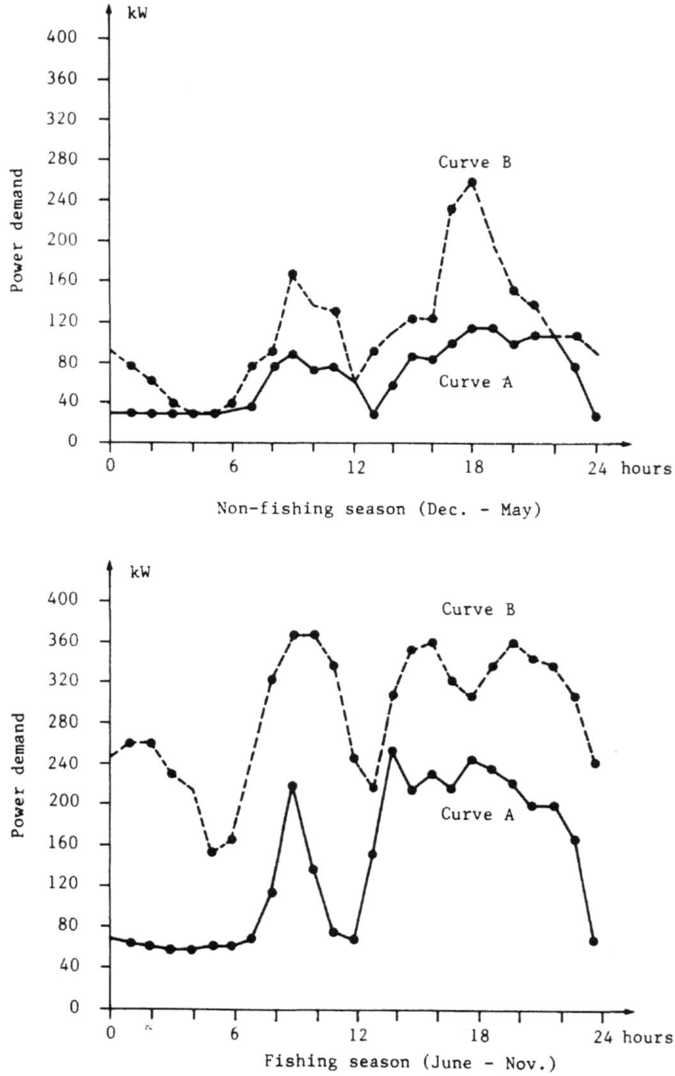

Xia Dachen, expected average load curves for 1988

Figure 4

Organization of Joint Wind-Diesel Group (JWDG)

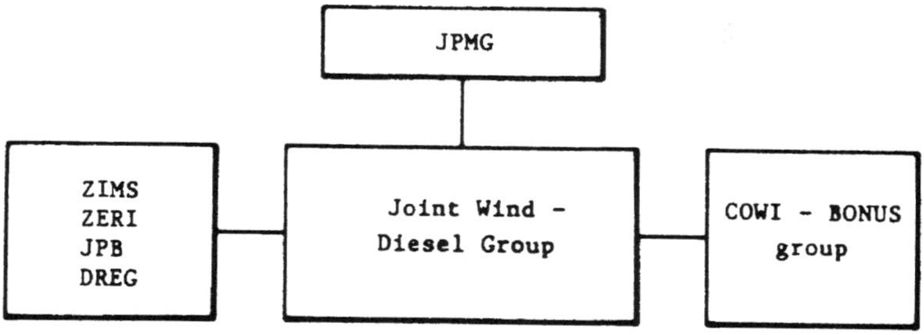

ZIMS : Zheijiang Institute of Mechanical Science

ZERI : Zheijiang Energy Research Institute

JPB : Jiaojiang Power Bureau

DREG : Dachen Renewable Energy Group

Figure 5

A SURVEY OF THE PAPERS PRESENTED

H Petersen
Wind Energy Consultant to the CEC (DGXVII)

1. INVITED PAPERS

As an introduction to the workshop, two invited papers were presented. Prof. N. Lipman gave a comprehensive survey of wind/diesel systems highlighting the different principles of system design, the combinations of wind turbine, diesel engine, storage alternatives and control philosophies. Dr. G. Cramer presented an overview of the control principles and techniques, discussing the methods to achieve optimum utilisation of wind/diesel systems.

2. THE PROJECT REPORTS

At the contractors' meeting, reports were given on seventeen projects. The projects can be divided into three categories: (a) wind/diesel systems designed for low penetration, wind turbines having a rated generator power of less than 10% of the installed power of the diesel generators; (b) wind/diesel systems for medium or high penetration, the ratio of the wind turbine rated power to the diesel generator power being mostly more than 50%; and (c) autonomous systems not supported by engine power, i.e. battery charging wind turbines.

In seven of the projects, all of them on islands, the wind turbine or turbines have a rated power which is less than 7% of the installed power of the diesel generators resulting in low penetrations. The projects are listed below.

Category (a) - Low Penetration					
Project	Manufac-turer or contractor	Location	WT power KW	Diesel power MW	Ratio WT/DG
127/83	VAWT	I. of Scilly	100	2	0.05
370/83	PPC	Mykonos	108	12	0.01
403/83	PPC	Karpathos	175	5	0.04
476/84	NSHEB	Shetland	750	30	0.03
619/84	Grenaa	Anholt	40	0.7	0.06
626/84	PPC/NTUA	Skyros	100	) 2.5	0.07
209/85	PPC/NTUA	Skyros	80	)	

Ten of the projects are on autonomous systems, seven of them being wind/diesel systems for medium or high penetration, while three are battery charging systems not supported by engine power.

Five of the wind/diesel systems are for the supply of electricity in general. Two of the wind/diesel systems are for special applications, one of them for aerobic treatment of sewage, the other for the pumping of water. Some characteristics of the projects are listed below:

Category (b) - Medium/high Penetration						
Project	Manufac-turer or contractor	Location	WT power KW	Diesel power MW	Ratio WT/DG	Battery capacity KWh
157/83	Aeritalia	Calabria	20	2x20	0.5	12
306/84	SMA Regel	Cape Clear	2x30	72	0.83	100
324/84	MBB	Lemwerder	20	25	0.8	0 (1)
091/85	CENG	Indian Ocean	30	120	0.25	0
147/85	Aeritalia	Montorio AIV	225	375	0.6	0
376/86	GESA	Palma de M.	30	2x20	0.75	-
405/86	ECOTECNIA	Castilla I.M	25	16	1.56	0 (2)

(1) Aerobic treatment of sewage
(2) Pumping of water

The remaining three projects are autonomous battery charging wind turbines not supported by engine power. They are located at three islands of the Glenan archipelago, Brittany, France. The generated electricity is used partly for residence supply, lighting, refrigeration, small tools, etc., and partly for special applications, at Penfret for pumping of water, at Saint Nicolas for water treatment in aquaculture plants and at Drenec for desalination of seawater.

Category (c) - Battery charging without engine power					
Project	Manufac-turer	Location	WT power KW	Battery capacity KWh	Applications besides general elec. supply
157/83	Aerowatt	Penfret	5	60	Pumping
306/84	Aerowatt	St Nicolas	5	-	Aquaculture
092/86	Aerowatt	Drenec	10	-	Desalination

3. A REVIEW OF THE PROJECTS

Generally the status of the projects is such that they are at the design or construction phase or in the phase of measurement data acquisition. Therefore, no final evaluations, especially relating to operation economies, are yet available.

As described in the foregoing, seven of the projects relate to wind turbines connected to a relatively strong grid (category (a)), the wind turbine rated power being 1 to 7% of the installed diesel generator power. Such applications are normally straightforward as far as grid connection is concerned, as the control system can be of standard design. The installation of a wind turbine on an island, however, can present problems with transportation and erection, and the maintenance costs of a single unit on an island can also be considerable. It will therefore be most interesting to examine the data when available and to see if there are ways in which costs can be reduced.

Another seven of the projects are on wind/diesel systems, five for the supply of electricity for general purposes and two for special purposes - the aerobic treatment of water and water pumping. With the exception of one project, the ratio of wind turbine rated power to diesel generator power is above 50%, and therefore designed for medium or high penetration (category (b)).

The projects differ very much in the philosophy behind the control system design - some are equipped with a battery storage, others have no energy storage. Several of the wind/diesel systems are now operational and data are being collected but to date very little is available for evaluation purposes.

For the future, it is believed that there is a large market for wind diesel systems. The market in Europe may be limited but in developing countries the demand for electricity is considerable, especially in small isolated places where the price of fuel is often extremely high. The problems will be to lower the price sufficiently to make wind/diesel systems competitive.

As is always the case, one way to lower the price is to produce the wind/diesel systems in large quantities and to market and distribute the systems in a rational manner. In some countries, the only way would be to produce the systems locally under licence. The diesel engines are often produced locally.

Most of the projects can be described as pre-commercial, featuring advanced control systems governing the operation by improved strategy. Such projects cannot be expected to be cost efficient for which reason it will be complicated to extract economic data directly from the results of operation. However, much experience will be gained from optimising the systems.

In order to reduce the fuel and maintenance costs of the diesel engine, it seems to be essential that the diesel should run at maximum efficiency during most of its running time and never run idle. The frequency of stops/starts should be kept low, say 10 times a day.

These conditions can only be met when storage systems are incorporated. Some of the projects incorporate a battery store, and it will be interesting to compare these with the systems without battery store and with the experience of wind/diesel systems with dump load control.

An interesting kind of storage not used in the projects is the flywheel. Systems utilising a flywheel for short-time store are being studied and applied in actual wind/diesel systems, and seem very promising.

The last three projects to be mentioned are the autonomous battery charging wind turbine systems. Basically, this technique is very old but is still undergoing development towards high efficiency and durability. The projects will further the achievements already made.

CONCLUDING REMARKS

K. Diamantaras
Wind energy Expert to the CEC (DGXVII)

In this second CEC contractors meeting, the participants exchanged useful information on technical and financial problems which they met in the course of their wind demonstration projects on autonomous and wind/diesel systems.

Numerous valuable and useful conclusions can be drawn from this meeting. These include:

1. Wind penetration could be as much as 30-35% in terms of installed power in a diesel-based grid without causing special problems.

2. Wind penetration, in terms of generated energy, could be as high as 50% leading to an almost equivalent amount of fuel savings.

3. In most applications a rather sophisticated load and control management is required, some times with the addition of dump loads. These latter ones should be as useful as possible so that it could be considered as a useful application and have a positive contribution as far as it concerns the economics of the project.

4. Applications which have an inherent character of storage, such as desalination, refrigeration, pumping, should be preferred compared to applications which require direct use of the wind energy.

5. Battery storage is still considered as very expensive and their use is advisable only if there is no other alternative.

6. There is, however, still much work to be done in the R, D & D field in order to achieve a deeper knowledge of these systems.

From an economic point of view, wind/diesel systems are in general competitive due to the fuel savings and the cost of transporting the fuel to the site.

Wind energy autonomous sytems require some means of energy storage which in most cases is very expensive and thus raises the total cost of the plant dramatically. However, depending on the alternative solutions, these systems may be competitive and important in social terms.

It has been emphasised that greater collaboration is needed between manufacturers and researchers in this wind energy field. Furthermore, standardisation of the wind-diesel system would lead to more reliable and cost-effective systems.

LIST OF PARTICIPANTS

ALEXOPOULOS, V.
PPC
Xalcocondili 22
GR - ATHENS

ANDERSON, G. A.
North of Scotland
Hydro Electric Board
Rothesay Terrace 16
UK - EDINBURGH

ANDREASEN, S-E.
COWICONSULT
Teknikerbyen 45
DK - 2830 VIRUM

ANDROUTSOS, A.
PPC/DEME
Navarinou Street 10
GR - ATHENS

ATHANASSIADIS, N.
National Technical University
of Athens (NUTA)
Patision Street 42
GR - ATHENS

BAKIS, E.
PPC
Navarinou Street 10
GR - ATHENS

BARBIERI, D.
Agence Française pour la
Maîtrise de l'Energie (AFME)
Route des Lucioles
F - 06565 VALBONNE CEDEX

BERGELES, G.
Lab. of Aerodynamics
Nat. Techn. University, Athens
Patission Street 42
GR - ATHENS

BETZIOS, G.
PPC
Parnasidos 38
GR - ATHENS

BLOTTO FINADRI, A.
AERITALIA
Via Archimède 156
I - ROME

CARATTI, G.
CEC - DG XII
rue de la Loi 200
B - 1049 BRUXELLES

CASALE, C. A.
ENEL
Centro di Ricerca Elettrica
Via Volta 1
I - 20093 COLOGNO MONZESE (MI)

CHADJIVASSILIADIS, J.
Public Power Corporation
Navarinou Street 10
GR - 10680 ATHENS

CHRISTENSEN, C. J.
Riso National Laboratory
Dept. of Meteorology and
Wind Energy
Frederiksborgvej 399
DK - 4000 ROSKILDE

COROMINAS, J.
ECOTECNIA
Demostenes 6
E - 08028 BARCELONA

CRAMER, G.
SMA-Regelsysteme GmbH
Hannoversche Strasse 3
D - 3501 NIESTETAL

DAVIS, M.
Commission of the
European Communities
rue de la Loi 200
B - 1049 BRUSSELS

DIAMANTARAS, K.
Commission of the
European Communities
Directorate General for
Energy
rue de la Loi 200
B - 1049 BRUXELLES

FINLAY, H.
Commission of the
European Communities
DG XVII
rue de la Loi 200
B - 1049 BRUSSELS

GAVRILIDIS, P.
Ministry of Energy
Michalakopoulou 80
GR - ATHENS

GEORGIOU, L.
PPC
Navarinou Street 10
GR - ATHENS

GEORGIOU, L.
PPC
Sigrou Street 112
GR - 117 41 ATHENS

GOUGE, H.
Territoire des Terres
Australes et Antarctiques
Françaises (T.A.A.F.)
rue des Renaudes 24
F - 75011 PARIS

GREBE, R.
SMA-Regelsysteme
Hannoversche Strasse 3
D - 3501 NIESTETAL

HUPPMANN, G.
Messerschmitt-Bölkow-Blohm GmbH
P.O.B. 801109
D - MUNCHEN

KALAGIA, H.
PPC
Navarinou 10
GR - ATHENS

KANELLOPULOS, D.
PPC/Alternative Energy Forum
Navarinou Street 10
GR - 106 80 ATHENS

KARANGELOS, J.
PPC
Navarinou Street 10
GR - 106 80 ATHENS

KLEINKAUF, W.
University of Kassel
Im Bodenfeld 18
D - 3500 KASSEL

KORONIDES, A.
PPC
Chalcocondyli Street 5
GR - ATHENS

KRAVARITIS, A.
PPC
Chalcocondyli Street 30
GR - 104 52 ATHENS

LIPMAN, N.
Rutherford Appleton Lab.
Chilton
UK - DIDCOT, OXFORDSHIRE

MATHIOULAKIS, D.
Public Power Company
Navarinou Street 10
GR - ATHENS

MAYS, I.
Vertical Axis Wind Turbines Ltd.
St. Albans Road 1
UK - HEMEL HEMPSTEAD, HERTS. HP2 4TA

MITSCHEL, H.
CONSULECTRA/HEW
Flotow Str. 41-43
D - 2000 HAMBURG 76

MORTENSEN, K.
ELKRAFT
Lautruphøj 5
DK - 2750 BALLERUP

NACFAIRE, H.
Commission of the
European Communities
rue de la Loi 200
B - 1049 BRUXELLES

NOEL, J-M.
AEROWATT-INTERNATIONAL
avenue des Coquelicots 6
F - 94385 BONNEUIL CEDEX

NOGARET, E.
AFME-Ecole des Mines France
rue Claude Dauresse
Sophia Antipolis
F - 06560 VALBONNE

PAPADOPOULOS, M.
National Tech. University of
Athens
Troados 27
Agia Paraskevi
GR - ATHENS

PAPAILIOU, K.
National Technical University
of Athens
Lab. of Thermal Turbomachines
P.O.B. 64069
GR - 157 10 ATHENS

PASCUAL TORTELLA, J.
Gas Y Electricidad S.A.
Juan Maragall 16
E - 07006 PALMA DE MALLORCA

PASSEY, D. J.
James Howden and Company Ltd.
Old Govan Road
Renfrew
UK - PA4 OXJ SCOTLAND

PETERSEN, H.
CEC-Consultant
Klostermarksvej 17
DK - 2700 BROENSHOEJ

PETRIDIS, A.
PYRKAL S.A.
Ilioupoleos 1
Ymittos
GR - ATHENS

SKAMRIS, C.
EUKRAFT A.m.b.A.
Lautruphøj 5
DK - 2750 BALLERUP

STELAKATOS, C.
PPC
Chalcocondyli 30
GR - ATHENS

TASSIOU, R.
Public Power Corporation
Agia Lavras 4
Pendeli
GR - 152 36 ATHENS

TSIPOURIDIS, J.
PPC/DEME
Navarinou Street 10
GR - 106 80 ATHENS

TRIGAZIS, S.
Public Power Company
Navarinou Street 10
GR - ATHENS

TZIROS, K.
PPC
Navarinou Street 10
GR - ATHENS

VALTER, G. P.
E.C.N.
Post Box 1
NL - 1755 ZG PETTEN

VAN DER STEEN, A.
Provinciaal Electriciteits-
Bedrijf van Noord-Holland
Ign. Bispincklaan 19
NL - 2061 EM BLOEMENDAAL

ZERVOS, A.
National Technical University
of Athens
Patission Street 42
GR - 106 82 ATHENS

ZOVANOS, N.
PPC/Island's Department
Singrou 112
GR - 117 41 ATHENS

INDEX OF AUTHORS